MG from A to Z

Jock Manby-Colegrave, a director of the Remenham Hill-based Squire Motors, at the wheel of his K3 Magnette, the greatest of all MG racers, prior to the 1933 Tourist Trophy.

MG from A to Z

Jonathan Wood

MOTOR RACING PUBLICATIONS LTD
Unit 6, The Pilton Estate, 46 Pitlake, Croydon CR0 3RY, England

First published 1998.

British Library Cataloguing in Publication Data

Wood, Jonathan
MG from A-Z
1. MG automobile – Encyclopedias
I. Title
629.2'222

ISBN 1899870296

Printed in Great Britain by MPG Books Ltd, Bodmin, Cornwall

Front cover: Gerry McGovern, who has the acclaimed lines of the MG*F* to his credit, at the wheel of a 1.8i in the model's 1995 introductory year. Inset: J2. Rear cover drawing by Alan Marsh.

About the Author

Jonathan Wood, The History Man of *MG World* magazine, is an MG enthusiast of long standing and as a founder member of the staff of *Classic Cars*, ran an MGA coupe as a company car. The author of over 30 books on motoring history, he has twice been awarded the Guild of Motoring Writers' Montagu Trophy and is a two-times recipient of the America-based Society of Automotive Historians' Cugnot Award.

MG 1434
Brockbank

Introduction and Acknowledgements

If I was looking for a subtitle for *MG From A to Z*, it would be MG People, Places and Things. The **people** are the managers, engineers, mechanics, drivers and stylists who have contributed so much to ensure that MG is one of the few British marque names to survive into the 21st century.

The **places** are those garages and factories that have played host to the make. Its origins are to be found in the City of Oxford, Abingdon-on-Thames was its home for 51 years and, finally, the Rover Group's Longbridge factory is where the acclaimed *MGF* is produced. For thankfully MG is, once again, a *living* marque.

But history is never static, and the old A Block of the MG's Abingdon Works was demolished whilst I was writing this book. However, the reader could spend a very pleasant and fulfilling day in Oxford and Abingdon visiting the MG sites, perhaps ending up at the *Boundary House* or *Magic Midget* pubs in the latter town for a glass of Old Speckled Hen ale, followed by a meal at the *Crown and Thistle* . . .

I have deliberately not written in great detail about MG cars, other than the famous record-breakers, because they have been so ably and exhaustively covered by other titles. These have been listed in the bibliography at the end of the book.

However, I hope that some of the **things** I have set down about MG, and the companies which supplied it, may be new to you, and some of the other entries may raise a smile or two. This is a book that can either be read through from A to Z or dipped into as a reference source.

Over the past 25 or so years I have talked with scores of MG people, some of whom, alas, are now dead. This was first as a founder staff member of *Classic Cars*, and subsequently as a motoring author and currently as The History Man of *MG World* magazine.

I owe a particular debt to John Thornley, MG's general manager from 1952 until 1969, who provided me with a fascinating insight into the marque and the indefinable Abingdon Spirit, established by Cecil Kimber, that he nurtured so carefully. Thanks are similarly due to Roy Brocklehurst, Jack Daniels, Syd Enever, Don Hayter, Ted Lee, Ted Lund and Henry Stone.

Over the years I have talked with British Motor Heritage's David Bishop, Rover Special Products' Graham Irvine, Stephen Schlemmer and Mark Gamble about the MG RV8.

Similarly, having interviewed Brian Griffin, Nick Fell and Gerry McGovern, I now know why the *MGF* has spearheaded the marque's 1995 revival so successfully.

I owe particular thanks to Robin Barraclough, co-author with Phil Jennings, of *Abingdon to Oxford*, the definitive and illuminating work on MG's shadowy origins. I am also grateful to Peter Seymour in this context. Brian Moylan put me right on some details of MG's Abingdon factory, whilst Ralph Liddiard provided fascinating information on the Pavlova Leather Company, which built what became MG's world-famous Abingdon Works.

What was once the MG office block is now named Larkhill House and is owned by the construction services company Barwick. My thanks to John Barwick for showing me around the recently renovated premises and giving me the opportunity to view what had once been Cecil Kimber's office and the MG boardrom above.

The MG Car Club's FWD Registrar, Roger Parker, helpfully supplied available chassis numbers for the MG Metro, Maestro and Montego.

I am also most grateful to Ian Cooling, author of *Jaguar Collectibles*, for supplying me with detailed information on his researches into the MG Tigress Mascot Mystery, which stretched back well over a decade.

The true identity of Barré Lyndon was a closed book to me until I spoke with *The Automobile*'s John Willis, the Surrey Performing Arts Library, the Theatre Museum and the British Film Institute.

Morland's Silvana Payne provided information and photographs of the Abingdon's brewery's involvement with the marque. The MG Car Club, MG Owners' Club and Octagon Car Club all supplied information relating to their activites. I am grateful, as ever, to Rover's Kevin Jones and Harry Edwards and Barney Sharratt, respectively of the Morris Register and the Austin A30-A35 Owners' Club.

The library staff of the National Motor Museum at Beaulieu, Annice Collett and Marie Tieche, offered ever-helpful assistance, as did their opposite numbers at Abingdon Library and the Centre for Oxfordshire Studies in Oxford.

You may have bought a J2 Midget back in the 1930s, or maybe are a well-satisfied MGB owner, perhaps you have just taken delivery of a new *MGF*, or indeed are just fascinated by the marque. But whatever your commitment to MG, I trust you will find much of interest within the pages that follow.

April 1998

Jonathan Wood
Frensham, Surrey

The BMC **A Series** engine that powered the MG Midget from 1961 until 1974, the 1100/1300 front-wheel-drive saloon and, ultimately, the MG Metro, was designed at the Austin factory at Longbridge *qv*. Work on the engine was well underway by 1949, the four-cylinder pushrod unit, coded AS3, being designed for Austin's A30 saloon of 1951. Whilst the work was undertaken to the brief of technical director John Rix, the layout and detail was undertaken by Eric Bareham *qv*, deputy to Bill Appleby *qv* in the engine division.

Much of the AS3's layout had been dictated by Austin's existing 1.2-litre 'four' that had appeared in the A40 of 1947. The new smaller unit, therefore, employed the same bore/stroke ratio of 1.3:1. With a bore of 57.9mm and a 76.2mm stroke, this helped to keep the engine's length to a minimum. Capacity was 803cc and the unit developed 28bhp at 4,800rpm.

Like its A40 predecessor, the side-mounted camshaft and pushrods were placed on the right-hand side of the engine. This meant that the carburettor and exhaust pipe were adjacent and well away from the electrical equipment, so obviating the risk of fire.

But this meant that the drive from the camshaft-driven distributor had to be taken across the engine and, to avoid the use of another costly camshaft-mounted gear, the oil pump was operated directly from the flywheel end of the camshaft. A three-bearing crankshaft was employed. At the other end of the engine the cylinder head incorporated Harry Weslake's distinctive heart-shaped combustion chambers, which promoted more efficient fuel burning and thus petrol consumption.

By the time of its appearance in 1951, Bareham had named the new unit the A Series, to differentiate it from the A40 'four' that soon afterwards he revised and which became the B Series *qv*.

In 1956 the A's capacity was increased to 948cc for the Austin A35 and Morris Minor 1000 saloons by enlarging the bore to 63 mm, and this version developed 34bhp.

As the result of ministrations by Eddie Maher *qv* at Morris' Engines

Branch *qv* it formed the basis of the unit used in the Austin-Healey Sprite of 1958, although it was fitted with twin SU carburettors and in this form output was raised to 42bhp. It thus also powered the MG Midget derivative of 1961.

An increase of the bore and stroke to 64 x 83mm, giving 1,098cc, for the Morris 1100 saloon of 1962, saw this engine also produced in twin-carburettor 55bhp form for its MG derivative, and simultaneously it was extended to the Midget and Sprite two-seaters.

The year 1967 brought the arrival of the Mark III Midget and Mark IV Sprite, with a further enlarged 70 x 81mm, 1,275cc unit courtesy of the Maher-developed Mini-Cooper S 'four', although detuned to 65bhp. The Austin-Healey version was discontinued in 1969, but this engine continued to power the Midget until 1974, when it was replaced by the more emissions-friendly 1,493cc Triumph Spitfire unit.

In the meantime, the one-time Austin 'four' was uprated in 998cc and 1,275cc A Plus forms for the 1980 Metro hatchback, and it therefore powered the MG Metro of 1982. But although the 1,275cc capacity was shared with the top of the range Austin version, the MG's engine was peculiar to the model, the 72bhp it produced being 12bhp more than the standard version's output. The compression ratio was raised from 9.4:1 to 10.5:1 and a new camshaft with an even wider overlap than in the revived Mini-Cooper engine featured. Enlarged inlet valves with modified throats were introduced, and the engine was identified by a ribbed and polished rocker cover.

In 1983 the Metro Turbo arrived, and the turbocharged version was boosted to 93bhp (DIN) at 6,150rpm. The ministrations, undertaken by Lotus, saw the relatively high compression ratio unexpectedly retained, but there was a sophisticated wastegate system on the Garrett T3 turbocharger that was mounted almost directly below the SU HIF 44 carburettor. Both models were discontinued in 1990, but the A Series, in fuel-injected form, is still being produced at Longbridge for the seemingly evergreen Mini at the rate of 350 to 400 a week. In production for 47 years at the time of writing (1998), it is Britain's longest-running car engine.

The arrival of MG was destined to make the name of the small county town of **Abingdon-on-Thames** world-famous. Hitherto the community, some six miles from Oxford, had been best known for its ruined Benedictine Abbey and magnificent 17th century town hall, designed by Christopher Kempster, a pupil of Sir Christopher Wren, as befitted the county town of Berkshire.

Industry had been mostly confined to the wool trade, and the town became an important cloth-making centre in the 14th and 15th centuries. Later came the spinning and weaving of hemp and flax, and the production of leather.

The town was largely bypassed by the excesses of the Victorian age, probably because its MP stood out against the Great Western Railway passing through the community, although a branch line to Abingdon was built in 1856. Perhaps for this reason, three years later *The Quarterly Review* described Abingdon as "a quiet melancholy old place", and in 1870 it ceded its position as county town to Reading. It remained in Berkshire until 1974 when boundary changes put it into Oxfordshire.

When MG arrived on the western outskirts of the town in 1929, the population stood at some 7,000, and the car company would become Abingdon's largest employer. But today you will look mostly in vain for any traces of MG in the town. Yet in the 1950s it was so identified with the marque that MG guru Wilson McComb *qv* had written of it as "a sleepy little Thames-side village where the cows leave octagonal hoof-marks."

Nowadays, the *Boundary House qv* and *Magic Midget qv* pubs are worth seeking out for their MG associations. Happily, the MG Car Club *qv* is in Cemetery Road, near the site of the factory. On the one-time Ashville Trading Estate nearby is Nuffield Way, leading to Eyston Way and Kimber Road, although these predate the plant's 1980 closure.

Captain George Eyston's name is immortalized in this road name near to the factory site.

The two-mile long **Abingdon branch line**, that joined the town with the main Reading-to-Oxford service at the village of Radley, survived well into the post-Second World War years, largely because of MG's use of the service. Built by the local Abingdon Railway Company in 1856, it was taken over by the Great Western Railway in 1904 and nationalized, along with the rest of the system, in 1948.

MG began to use the line in earnest, although not for cars, during the war and it came into its own in the export drive of the 1940s. Until 1965 the trains were hauled by an ex-GWR tank engine but this was then replaced by a more anonymous diesel. By 1969, traffic had increased sufficiently for three trains a week to leave the town, with from six to ten wagons per train, each carrying four MGBs or Midgets for, ultimately, overseas destinations.

A team of some eight drivers were regularly employed at the factory to deliver cars to the station. As volumes increased, loading ramps were introduced there in 1973 to permit the use of two-tier wagons and there were 20 Carflat units in all. But with MG's closure in 1980, the last car-bearing train left on May 23 that year, traffic was dramatically reduced as a result and the line closed in 1984, the final journey being made on March 27. Today, all trace of Abingdon Station has disappeared.

A derelict clothing factory in West St Helen Street, **Abingdon**, was used to house MG's service and spares department between 1940 and mid-1948. The facility continued to function throughout war, but there was a fire there in the summer of 1944 and much material was water-damaged. Sadly, these were mostly spares relating to the MG racing cars and record-breakers that were stored in the basement. Instead of retrieving the bronze cylinder heads, superchargers and twin-cam conversions, it was decided to leave them there and the basement where they were stored filled with concrete.

Although the service department reverted to the factory in June 1948, the MG Social Club continued in the premises for a further 10 years. In 1958, when MG was given some land by Abingdon council in exchange for the old store, the building was demolished. Today, the site is occupied by a block of flats.

The limited-edition **MG*F* Abingdon,** announced at the 1997 Motor Show, was restricted to 500 units, 150 of the entry-level model and 350 with the VVC (Variable Valve Control) engine. Finished in British Racing Green, the cars had traditional leather upholstery and a walnut dashboard. Deliveries began in March 1998.

The **'Abingdon Pillow'** was the nickname given by the MG factory to the energy-absorbing dashboard (created in deference to American safety regulations) that faced passengers in US-sold MGBs. It applied from 1968 until 1971.

The *MGF* **Abingdon Trophy,** launched in 1998 by Rover's MG Cars *qv* subsidiary, is a contest which has been designed for owners wishing to compete in their own road cars at minimum cost. A 12-round competition is administered by the MG Car Club *qv*, and registered competitors can avail themselves of a subsidized motorsport race kit.

MG's home for 51 years, from 1929 until 1980, the **Abingdon Works** was so named after the Second World War and followed its purchase, in 1940, by Morris Motors *qv*. Previously it had been known as the Pavlova Works, having been built by the Pavlova Leather Company *qv* as an extension to its own factory during the First World War.

When MG arrived there, in September 1929, the building had remained unused for some years, although The Morris Garages *qv* in Oxford was always running out of space, and rented it to store used cars. This required visits by members of the Garages' workforce, and it was one of these mechanics, Sydney Enever *qv*, who became familiar with the works. In 1929, when MG was outstripping the confines of its purpose-built Edmund Road factory in Cowley *qv*, Enever suggested that the vacant Pavlova premises might make an ideal car factory.

The Leather Company was approached by Morris, who agreed to a five-year lease for an annual rental of £1,250. The lease was renewed in 1934, but not in 1939 because by then Morris had begun negotiations to buy the works. On November 18 that year, Morris paid a deposit of £2,331 on the purchase, with the balance following on February 13, 1940, making a total cost of £23,315.

The Morris Owner in 1932 described it as "a super factory". Access was by Cemetery Road, a turning off Spring Road that, in turn, connected with the main Marcham Road. There was an administrative block on the left, and Cecil Kimber installed himself in the first-floor office at the west end of the building and introduced a new bow window for his room and the boardroom above.

When a representative of *The Autocar* visited Abingdon in 1934, he noted that Kimber had "a most unusual office; it is in Tudor style, done by his own workpeople, simple but in excellent taste. One window looks out [to the south] over an orchard, and another to the traffic of MGs as they leave the works".

Kimber had the factory's interior brickwork painted in the MG colours, with the lower section finished in brown and the upper portion cream. On the occasion of an earlier visit, in 1932, by that

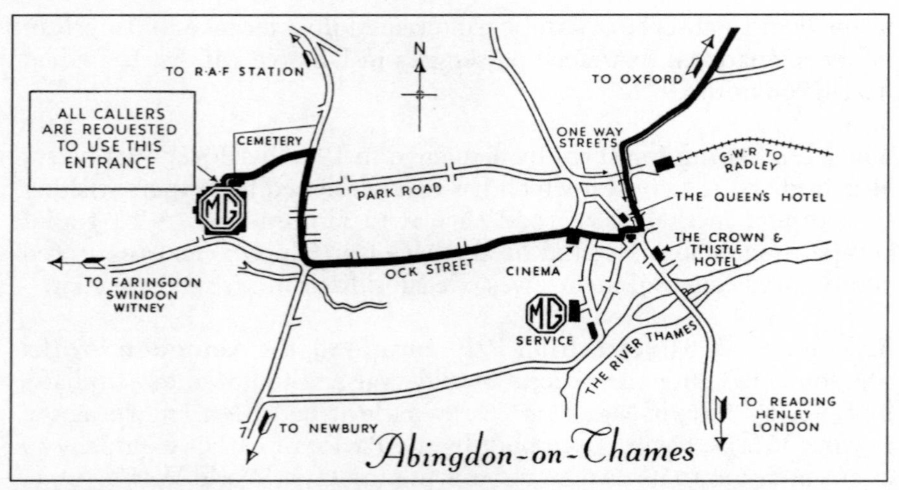

The location of the MG factory and then separate Service depot, as depicted in the early postwar years.

journal's Maurice Sampson, he was impressed by the "beautifully equipped, and marvellously clean factory . . . The first impression was one of spaciousness. No-one is cramped; everybody had ample elbow room. The whole atmosphere is keen, alert and enthusiastic. It is the sports car atmosphere." He had noticed the same air at the Bugatti factory at Molsheim.

After the administrative block, the road divided, right to the Pavlova Leather Company and left into the MG works. First visitors would encounter a glass-covered yard where cars were washed prior to dispatch. This was the entrance to the spotlessly clean boiler room, which doubled as a laundry where employees' overalls were washed on a weekly basis with 3d (1p) deducted from their wages for the service.

Beyond was the entrance to the works that, throughout its 51 years, was essentially an assembly plant. The chassis were always pushed along manually, just as they had been at Morris' Cowley factory until its 1934 modernization when powered assembly lines were introduced.

Inside was the office of works manager George 'Pop' Propert, who had joined Kimber in that capacity in 1925 and, after a brief spell with Bean Cars, returned to Edmund Road and remained with MG until his retirement in 1949.

The partially two-storey factory was divided into nine bays and its layout varied in detail, if not overall layout, up until the outbreak of the Second World War in 1939. One significant change occurred in the mid-1930s when the drawing office was transferred to Cowley. A paint shop, complete with drying oven, was also introduced.

On the ground floor, next to Propert's office, was the experimental department, with the engine rebuild and test shop next to it, overseen by Syd Enever. A machine shop was located alongside. Half of the storey above was occupied by the drawing office and the other portion by the factory maintenance department.

Chassis assembly and wiring was undertaken in bays 2 and 3 and in the latter was the ramp that led up to the body deck. Bodies were delivered from the Coventry-based Carbodies *qv* and from 1935 Morris' Bodies Branch *qv*, although the wings were painted at Abingdon. Before the creation of a figure-of-eight test track, that used cinders from the boiler house, the cars were evaluated by being driven round inside the building in bay 4, which was also used to rectify any problems that arose.

Assembly lines occupied bays 5 and 6. The stores also ran the length of the bay. The latter was initially used in 1930 for producing the Mark I 18/80 and Midget. But following MG's takeover by Morris Motors, two additional lines were introduced in 1936, although they occupied the same space.

Bay 7 was occupied by the service stores and service repair shop with the body deck above. In bay 8 was the service workshop, dispatch department and service offices. Bay 9 was initially sublet to the fizzy drinks manufacturer Schweppes; a quarter of the bay was used by the service department with the balance reserved for new car dispatch. Service customers entered by the entrance facing Marcham Road. However, completed cars being taken through the factory went out through the main gate in Cemetery Road.

The Pavlova Works in the 1930s, as drawn by Frederick Gordon Crosby's equally talented son Peter

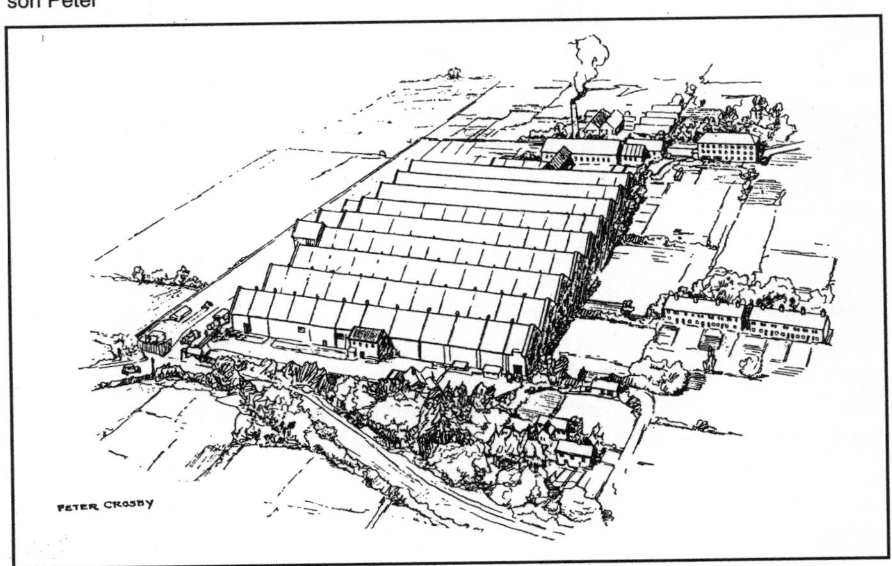

The MGB was the last MG to be built at Abingdon. This picture dates from 1980.

During the war three new buildings were erected, at the government's expense, on land to the west of the factory, which was the start of what was to become B Block. At this point the original works became A Block. B1, B3 and B5 were cannily sited by the factory's maintenance foreman, Jack Lowndes, by eye so when the war was over it was possible to fill in the gaps and produce more buildings. B6 was accordingly completed in 1950, B4 followed in 1952, B7 and B8 dated from 1954 and B2 in 1956. B0 followed in the late 1960s.

When the service department moved back to the factory from West St Helen Street *qv* in 1948, it occupied B1 building and continued so to operate until 1964. Its original location in A Block was used as a service department stores until it closed in 1964. The space was briefly occupied by BMC Special Tuning *qv* and subsequently a Sprite production line was laid down.

By the 1960s B Block variously housed, from the Marcham Road, the tyre fitting bay, competition department, dispatch, final inspection, paint repair ovens, finishing and rectification shops.The pre-1939 Marcham Road entrance was greatly enlarged during the war, and after production had been restarted it replaced Cemetery Road as the main factory access, although the latter continued to be used by the MG workforce.

In 1965 C Block was built to the north-west of B Block and occupied by the BMC Competitions Department *qv*, transport and stores. Soon afterwards the APCC (Air Pollution Control Centre) was

The compound in 1980 with Limited Edition MGB GTs awaiting dispatch.

constructed just to the south of it, which handled not only MG's emissions legislation but also catered for all British Leyland vehicles.

When the MG factory closed in 1980 it occupied 54 acres and employed 1,100 people. A few buildings were then demolished, most notably what was known as the Pavlova Engine House, attached to A Block, and used for a time by the development department. It came down in 1981, along with C Block, but A Block remained intact and B was retained but reclad. However, A Block was demolished in 1997 and Abingdon's new police station is to be built on the site.

The **Abingdon Works** was built on the site of a Roman christian cemetery. When the car park for C Block was being excavated in 1974, eight skeletons, some 1,500 years old, were uncovered. Seven were aligned east/west, but one was decapitated with its skull lying on the pelvis. Archaeologists believed it was the body of a criminal or witch . . .

ADO numbers were accorded to projects approved or initiated by the British Motor Corporation *qv* of the 1952-68 era. It survived into the British Leyland years, which came to an end in 1975 when the company was nationalized. When an MG project originated from Abingdon it was accorded an EX number *qv*; the MGB for instance was EX 214. But when it received corporate funding which could lead to production, it was also given an ADO number. In the B's case it was

ADO 23. They were allotted at random and not in numerical sequence. The MGA 1600, for instance, that preceded the B was ADO 31! As the Austin's Longbridge factory was BMC's design centre, logically the ADO prefix stood for Austin Drawing Office, but the merger with Morris required that the first letter stand for *Amalgamated.*

In 1938-39 MG **advertising** was handled by W S Crawford Ltd, the prestigious London agency. It was headed by Sir William Crawford, who was also chairman of the Buy British campaign, an initiative close to Lord Nuffield's *qv* heart. This is why some MG advertisements bore the legend 'British Cars are Better Built - British cars last longer.'

Others carried a personal statement from Cecil Kimber, thus associating him in the public mind with MG which, apparently, greatly displeased Lord Nuffield. According to William Howlett, chairman and managing director of piston ring suppliers Wellworthy Ltd, who kept his ear close to the ground, the Morris chairman was "displeased when an Australian, misled by the promotion, refused to believe that the car was anything to do with Morris". The man told Lord Nuffield to his face that MG "was too much of a thoroughbred to have come out of his stables . . ."

The **advertising slogan** that MG used to launch the VA model of 1936, 'For space . . . for grace . . . for pace', was 'borrowed' after the Second World War by Jaguar for its commodious Mark VII saloon of 1951. However, it rearranged the message to read: 'grace . . . space . . . pace'. In mitigation to Jaguar, in 1935 its SS predecessor was proclaiming its sports saloon's 'Speed . . . grace . . . silence', so perhaps that is where MG got the idea?

H W Allingham was responsible for the design of one of the most sought after MGs of the inter-war years, the factory-approved Airline coupe that appeared in 1934. Allingham was a co-founder, in 1921, with H Hamilton Hoyer and Chalmer, of Chalmer & Hoyer, which built Hoyal bodies. The firm was a victim of the recession and closed in 1931, but Allingham set up in business on his own account as a body designer with his work being produced by other coachbuilders. Based at 10 Stratford Place, London W1, his distinctive and elegant designs were taken up by Austin, Ford, Morris, Vauxhall and Wolseley in addition to MG.

The coupe, with its sliding roof containing three 'windowlets', two-tone livery and sliding windows, was constructed by Carbodies and offered on the PA chassis at £290, which was £68 more than the open two-seater. Just 28 were built on the P-type frame. It was also available on the N-type Magnette, although this was a slower seller, a mere six being sold. There were a further 14 examples on the PB chassis, with the remaining two used on the TA. One found its way on to an HRG chassis in 1938.

Allingham also designed the Two-Four Seater tourer for the N-type in which two additional passengers could be carried in a dickey seat that was enclosed by the hood. It thus had echoes of Kimber's Morris Garages Chummy *qv*. Just 11 examples were built by Whittingham and Mitchel, of London SW6.

The **Alvis-MG** hybrid is a rare car. After the Second World War coachbuilder Richard Mead of Dorridge wanted to body the Alvis TA 14 chassis and chanced to see a newspaper advertisement by Tickford *qv* for a batch of MG WA coupe bodies. During the war these had been consigned to a barn located behind a Newport Pagnell pub, and MG having decided not to continue with the WA, Tickford put them up for sale. Mead bought one and found that it fitted the Alvis perfectly and the finished product was illustrated in *The Autocar* in January 1948 with Mead credited with its construction!

As a result he bought all the remaining bodies, some of which were partially completed, and was able to produce 12 further complete cars. The majority went on Alvis chassis, but as they came with wings they lacked the elegance of MG's running-boards and side-mounted spare wheel. At least one was fitted to an 8-litre Bentley chassis. Of the dozen Mead produced, three are known to survive. He went on to body some of the rare Rover P4-based Marauders of 1952-54 vintage; he died in 1994.

America's first MG owner may well have been Henry Ford's son Edsel, who in 1930 bought an M-type MG (2-812). The purchase was arranged, on February 18, through the Ford Motor Company's London office. Edsel drove the car in and around Detroit for three years, by which time it had 27,509 miles on the mileometer. He subsequently donated it to the Henry Ford Museum at Dearborn, Michigan, which retained it for many years, although it is now displayed in a Pennsylvania museum.

Although the name of **William (Bill) Victor Appleby** (1903-1984) is associated more with BMC's A and B Series engines, in an MG context he is best remembered for designing a rare twin-overhead-camshaft engine that momentarily appeared in the sports-racing MGA in 1955. He was well qualified for the job.

Liverpool born Appleby was educated at Wolverhampton Grammar

School and studied engineering at Wolverhampton Technical School. After gaining practical experience in the machine shop of ABC Coupler in 1918, he gravitated to Sunbeam, the town's most famous employer, the following year. Once established in the drawing office, in 1921 he moved to the experimental department, where he worked under chief engineer Capt John Irving, who was in charge of preparing and developing the firm's twin-cam racing cars.

Bill Appleby left Sunbeam in 1925 and moved, successively, to Vauxhall, Bean and Thornycroft before joining Austin in 1934. He was recruited, along with Tom Brown, for the express purpose of working on the engine design of the twin-overhead-camshaft racing cars being developed under the direction of Murray Jamieson. Lord Austin's intention was to trounce the single-cam MGs in 750cc racing.

Despite opposition, Jamieson insisted on adopting a 100-degree valve angle rather than a narrower one, as favoured on motorcycle engines. This constraint is considered to have prevented these cars from enjoying greater success than they did.

When the programme came to an end in 1937, Appleby switched to production cars, although in 1942 he became section leader of engine design. His involvement in the design of the A40 and A Series units *qv* was one of overall responsibility, the work being undertaken respectively by John Rix and Eric Bareham *qv*. In 1952 Appleby became superintendent of engine design at Longbridge, and the creation of BMC *qv* saw him promoted to corporate engine and gearbox supremo in 1953.

Although Morris Engines had produced a twin-cam version of the B Series unit, at Len Lord's *qv* behest Appleby embarked on an all new design with a valve angle of 66 degrees, as opposed to the 80 degrees of the existing 'four'.

It was to have been used in one of the EX 182 MGAs being run in the 1955 Tourist Trophy race, but the night before the cars were due to leave Abingdon, Alec Hounslow *qv* received a late night phone call from Syd Enever *qv* to remove the unit from the car, whereupon it was crated and returned to Longbridge, never to resurface. Appleby retired in 1969 after BMC was absorbed by the British Leyland Motor Corporation *qv*.

The **Arnolt-MG** was the inspiration of Stanley Harold 'Wacky' Arnolt and was a Bertone-bodied TD produced in 1952 and 1953. Born in Albany, Georgia, USA, after graduating from Wisconsin University with an engineering degree, he obtained the manufacturing rights for a small marine engine and during the war secured a lucrative contract to supply them to the US navy.

With the coming of peace Arnolt built up a large Chicago-based commercial empire which ranged from a Mexico coconut plantation to newspapers and car accessories distribution. He bought an MG TC and in 1949 established S H Arnolt Inc, with premises at 415 East Erie Street, Chicago. He had a penchant for British cars and, in addition to MGs, also sold the Aston Martin and Bentley makes.

During a visit to the 1952 Turin motor show Arnolt met Nicole Bertone, head of the locally based Carrozzeria Bertone. Although established in pre-First World War days, by the early 1950s it had little work and was clearly in need of both orders and money. Bertone had purchased from MG two TD chassis, which they clothed with their own coupe and convertible bodies in the latest full-width Italian style that was much in evidence at the Turin event.

On seeing them, Arnolt ordered 100 and became vice-president of Bertone. The outcome was an Anglo-Italian hybrid, the Arnolt-MG of 1953, that became available in open and closed forms, but with two-plus-two seating. It was thus advertised as 'the new Arnolt Family-and-Sports car' and offered the Italian look, but without the usual mechanical complication. The selling price was initially $2,995, but rose to an uncompetitive $3,585.

Between November 1952 and May 1953 MG dispatched 100 TD chassis to Bertone, who bodied 65 coupes and 35 roadsters. But they were slow sellers, the last not being purchased until 1958-59. In the meantime, Arnolt had repeated the exercise on the Bristol 404.

Arnolt's association with MG drew him into contact with the British Motor Corporation, and he was best remembered for arranging a reputedly unsecured $1 million bank loan to purchase 1,000 Morris Minors, which was one of the largest export orders ever received by BMC. He became the corporation's distributor for the American North-West.

David Ash was co-driver with Phil Hill and Tommy Wisdom in EX 179's *qv* successful 12-hour record run at Bonneville in 1956. Wholesale sales director of J S Inskip *qv*, of New York, for 14 years, Ash reckoned to have sold some 5,000 MGs and to have set up approaching 200 dealerships.

As competent on the race tracks as in the showrooms, in the early 1950s David Ash built the MG Cigar Special, with independent rear suspension courtesy of a Mercedes-Benz 170, and successfully competed at Watkins Glen and elsewhere. Later came Motto, another MG special with tubular chassis, streamlined body and a factory engine prepared by Syd Enever. Later he competed in a TD Special. He was North American MG team captain and the only driver to start and finish the Sebring 12-hour endurance race on five separate occasions. Ash subsequently became a motoring writer and editor and a regular contributor to *MG Magazine qv*.

MG's great racing rival of the early 1930s, **Austin** was used to dominating the 750cc class with its legendary Seven. Introduced in 1922, this immortal side-valve-engined car soon established its competitive credentials in 1923, although from 1931 its supremacy was successfully challenged by the C-type Montlhery Midget *qv* with its single-overhead-camshaft engine.

Austin could be in little doubt of MG's ambition. Soon after the C's introduction, as R J Wyatt recounts in *The Austin Seven*, Captain Arthur Waite, who was Sir Herbert Austin's son-in-law and in charge of the racing programme, met Cecil Kimber at an Oxfordshire hotel. Kimber told Waite: "We are now at war. We hope it will be a friendly war, but it will certainly be war." The Captain responded: "I accept the challenge. My firm will fight. We will meet you on the racing tracks and in road races, and may the best man win."

The supercharged overhead-camshaft Midget soon underlined its superiority to the robust but ageing side-valve Seven. It was a contest that led, in 1934, to Austin appointing the talented young Murray Jamieson to design a completely new single-seater racing car powered by a twin-overhead-camshaft engine that bore little resemblance to the production Seven. Lord Austin committed an unprecedented £50,000 to the project, but by the time the cars appeared in 1936 MG had withdrawn from racing.

Then, in 1952, the Austin and Morris companies merged to create the British Motor Corporation, Austin becoming the dominant marque of the alliance. But its marketing strength was seriously undermined in 1974 following the arrival of the mechanically flawed and ungainly Allegro, a decline that culminated in the make's extinction when, in 1987, the name was removed from the Maestro hatchback and Montego saloon. The latter was the last new model to bear the Austin name.

Austin Abingdon was the name suggested, in 1955, by John Thornley *qv* for a tuned version of the Austin A90 saloon. This followed Ken Wharton having successfully campaigned the model in circuit racing. But the idea did not find favour with BMC's George Harriman *qv*!

Austin-Healey became the British Motor Corporation's corporate sports car when chairman Leonard Lord *qv* announced it in favour of alternative designs from MG's EX 175 *qv*, Jensen *qv* and Frazer Nash. The Longbridge-built Austin-Healey 100 evolved into the 2.6-litre 100-Six, powered by BMC's C Series *qv* engine. Manufacture was transferred to MG's Abingdon factory in November 1958, and the following year the capacity was increased to 3 litres and the car renamed the 3000. It continued to be built there in Mark II and III guises, the last example being produced on December 21, 1967, the month prior to the creation of British Leyland. Of the 72,022 'Big Healeys' built, a total of 51,315 cars, or 71 per cent, were manufactured by MG. A second, smaller model, the Austin-Healey Sprite of 1958, formed the basis of the MG Midget *qv* of 1961-79.

Some MGs were exported to **Australia** in chassis form in the 1930s because completed cars were liable to a higher tax to protect the local coachbuilding business. One such firm was C F S Aspinall, of Armadale, a Melbourne suburb, which undertook the work at the behest of Lane's Motors, MG's Australian agent. The bodies superficially resembled their British equivalents, although they were heavier. The N-type Magnette and PA were amongst the chassis to be so enhanced, but the tax was repealed in the mid-1930s and the practice gradually ceased, although a few SAs were locally bodied.

After America, **Australia** was the most popular market for the TC sports car in the early postwar years. A total of 1,774 examples were dispatched Down Under between 1945 and 1949, whereas the US took only 46 more, importing 1,820 TCs in the 1946-49 era. It was not until the arrival of its TD successor that transatlantic sales truly surged ahead.

The weekly magazine ***Autosport,*** founded in 1950 by MG enthusiast Gregor Grant (1910-1969), contributed no less than three personalities to MG's postwar history. They were George Phillips, Wilson McComb and the MG Car Club's Russell Lowry (all *qv*). Grant drove an MG Magnette in the 1956 Monte Carlo Rally and was placed eighth in class.

B

The BMC **B Series** engine that powered the MG Magnette saloon, MGA and ultimately the MGB appeared in its original form in the Austin A40 saloon of 1947. The origins of its layout are to be found in that company's first overhead-valve car engine, the 2.2-litre 'four' that appeared in the 16hp model announced in 1944.

This had been designed by John Rix in 1943, when Austin was asked to produce an alternative engine for the Jeep. He achieved it by lopping two cylinders off an existing Austin six-cylinder truck unit, itself inspired by a prewar Bedford truck engine!

As such a distinctive feature was the right-positioned camshaft, which meant that the carburettor and exhaust manifold were on the same side, well away from the electrical equipment in the form of the distributor, dynamo and starter motor on the left, this feature was bequeathed to the subsequent A and B Series units.

In addition to powering the 16hp, it found a more successful application under the bonnet of the 1948 Austin FX3 taxi, in bigger-bore 2.6-litre form in the ill-fated Atlantic saloon of the following year and, more significantly, the Austin-Healey 100 *qv*.

What became the A40 unit was also designed by Rix, soon to be Austin's technical director. As such it outwardly resembled a smaller version of the 16hp 'four', but differed in detail. Initially it was created in two capacities, of 1,000cc and 1,200cc, although eventually it was only produced in its larger-capacity 65 x 88mm form that featured a bore/stroke ratio of 1.35:1.

It inherited from the 16hp, so equipped from April 1946, the fuel-efficient distinctive heart-shaped combustion chambers created by Harry Weslake. As announced it developed 40bhp at 4,300rpm. The engine performed well in the A40, which was the most popular Austin of its day, and nearly 350,000 had been produced before the model was revised in 1952.

Following the creation of the BMC it was decided to uprate the engine because 1,200cc represented its maximum enlargement. The work was undertaken by Eric Bareham *qv*, who had already worked on the 16hp with Rix and had been largely responsible for the smaller A Series 'four'; the B designation logically followed on from it.

The intention was to produce a universal engine with the widest possible range of applications and the A40 unit represented the starting point. From the outset it was intended to manufacture it in 1500 and 1200 capacities but, to permit this enlargement, the block had to be extended and to reduce its overall length the water pump was recessed into the casting.

A new and stronger three-bearing crankshaft was required and the original white metal was replaced by much harder lead indium bearings which, in turn, demanded a redesigned oil delivery system. The B Series also had new bore centres, in accordance with new manufacturing processes, but the original crank-to-camshaft centres were preserved, along with the crank throw.

The heavily revised but outwardly similar engine first appeared in the MG Magnette *qv* of 1953 in 1,489cc form, achieved by increasing the bore size from 65 to 73mm, and it developed 60bhp at 4,600rpm. The 1,200cc cubic capacity, which retained the original internal dimensions, was perpetuated.

In this form the 'four' was used in twin-carburettor 68bhp guise for the MGA, although the bore was enlarged to 75mm for the B-based 1,588cc MGA Twin Cam, where power soared to 108bhp, although this was developed by Morris' Engine Branch *qv*. This capacity was extended to the 80bhp pushrod 'four' in 1959. In 1961 came a further bore enlargement, to 76mm, giving 1,622cc and 93bhp, and this MGA 1600 MKII was the first BMC recipient of the unit.

It was used in the prototype MGB, but luckily for Abingdon, at Longbridge the projected Austin 1800 saloon (ADO 17) of 1964 required that the 'four' be increased to 1,798cc and the 80mm bore represented the final enlargement of the B Series in a production car. In this form it developed 92bhp at 5,300rpm. With the arrival of the 1800, a five-bearing crankshaft was introduced which was also extended to the MGB.

When the MGB ceased manufacture in 1980, it was the only BL Cars model to be powered by the ageing unit. The intention was to replace it with the related O Series *qv* engine although this never reached the MG and gave the management of the day an additional excuse to discontinue the model.

The legless air ace **Douglas Bader** ran an MG J2 (J.2296) in prewar days that was specially converted for him to drive. After the war he immediately put in an order for a TC and it was duly delivered. But because he had more strength in his left stump than the right one the clutch and brake pedals were transposed. When he returned the car to

Abingdon for servicing, an MG employee, Tommy Boggs, who was moving it from reception, queried how to operate the controls, Bader responded: "Just cross your legs, dear boy!"

The famous MG octagonal radiator **badge**, that first appeared on the 14/40 Mark IV *qv* for the 1928 season, featured chocolate coloured initials set against a cream background. Although never officially confirmed, I like to think that Cecil Kimber was inspired by the livery of the Great Western Railway trains that served Oxford. In 1922, the year after Kimber arrived in the city, it had reintroduced this livery of the 1864-1909 era to its carriages after experimenting with crimson lake-hued rolling stock.

The design, the work of The Morris Garage's cost accountant Edmund (Ted) Lee *qv*, was initially used for that business' advertisements in 1923 and registered as a trademark in May of the following year. A keen amateur artist, in 1981 Lee told me that he had drawn it "with a little ruler I'd bought from [Oxford] High School". He showed it to Kimber, who said "that's just the thing," whilst William Morris maintained that "it was the best thing to come into the company," adding "and it will never go out of it." Lee said his choice of an eight-sided figure was of no significance, but its angular design and lettering is typical of what is currently known as the Art Deco style.

The enamelled badge was continued on MG's sports car line until the MGA was discontinued in 1962. That year it was revised for the

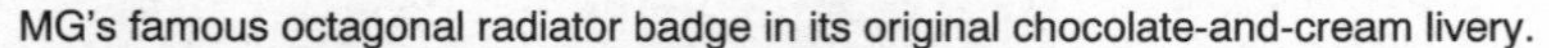

MG's famous octagonal radiator badge in its original chocolate-and-cream livery.

MGB and this cheaper version had silver letters and surround set against a red background. It was mounted on a black shield, and this design continued until 1969. That year the B's front end acquired a corporate British Leyland look with a recessed grille that required a new badge, and although the red and silver colour scheme continued it was now mounted on a black octagonal surround. With the reversion, for 1973, of a traditional grille, albeit a plastic one, the shield returned, but with a red rather than a black background.

The arrival of the 'rubber'-bumpered cars in the 1975 season required another change; the badge had a red background and the lettering was gold in the so called jubilee year of 1975. It then went to silver with a black background until the model ceased production in 1980. However, when the MG Metro appeared in 1982, its badge had silver lettering with a red background that was similarly applied to the Maestro and Montego variants. The 6R4 rally car had a smaller rendering with the same colour scheme.

When Rover Special Products was planning the MG RV8, it decided to redesign the MG badge and re-introduce the original pre-1962 chocolate and cream livery. This was duly unveiled in 1992, was perpetuated on the MG*F* of 1995, and now graces all Rover's MG literature.

One of the British motor industry's unsung heroes, **Eric George Bareham** (born 1915) was the Longbridge-based engineer responsible for the BMC A and B Series engines (both *qv*) used extensively by MG. Educated in London and Birmingham, in 1929 he joined Wolseley's Drews Lane works, following an interview with A V Oak, later chief engineer of Morris Motors, and worked in the drawing office for five years. In 1934 he moved south to Lagonda's service department, but it soon went into receivership and Bareham switched to Alta at nearby Surbiton, where he remained until 1938. There he undertook commissions for Harry Weslake, who was subsequently responsible for the combustion chambers of the Austin engines on which he was later to work.

A brief spell in 1938-39 with Electricars in Birmingham presaged his arrival at Longbridge in 1940. Later, in 1945-47, he had another spell at Lagonda before his return to Austin. On the retirement of his boss Bill Appleby *qv* in 1969 he replaced him as British Leyland's manager of its small car future projects division. Eric Bareham retired in 1974 to work for Noel Penny Turbines but briefly returned to Longbridge in a consultancy role for the third and final time in 1983 before a well-earned retirement.

The **best MG year,** in productive terms, was 1972, when 55,639 cars left Abingdon. This coincided with the annual record for MGB manufacture with 39,393 built; there were 16,243 Midgets and just three examples of the MGB GT V8.

The **best-selling MG** in the history of the make was the MGB, of which 512,243 examples were produced. Of these, 386,961 were roadsters and 125,282 GTs; some 83 per cent were exported, mostly to the USA. The MGB accounted for 53 per cent of all MGs assembled at Abingdon between 1929 and 1980.

The **best-selling MG-badged saloon** was not built at Abingdon but at Longbridge. A total of 120,197 MG Metro 1300s were produced between 1982 and 1990.

David Bishop (born 1944) was the instigator of the revived MGB bodyshell and the MG RV8 *qv* that sprang from it. He joined the British Motor Industry Heritage Trust in 1983, as assistant to managing director Peter Mitchell, from Pressed Steel *qv* at Cowley, where he had 23 years experience of body engineering. Ironically it was a request to save the body tooling for the MGB's Triumph TR7 rival in 1986, that proved impractical, which triggered the idea for the Trust's British Motor Heritage (q.v) subsidiary to remanufacture the B's roadster body. It was the first of a number of similar projects, and Bishop became its executive director in 1988. This in turn led to the RV8 of 1992 with an MGB-based hull, although David's idea was always for a cheaper car than the one that eventually emerged. He left the Heritage organization in mid-1995.

'Bitsy' was a short-wheelbase tractor created at the MG factory in 1939 because all of its lorries had been requisitioned by the army. It was used to transport car spares from the works to the stores in West St Helen Street, Abingdon *qv*, and so named because of being made of various bits and pieces and was even accorded in own EX 167 number. Built by the experimental department, it consisted of a cut down K2 chassis, flat-fronted MG 14/40 radiator and XPAG 1.2-litre TB engine. Gear ratios were sufficiently low for Bitsy to pull eight tons and three trailers. It proved to be so useful that it remained in use long after the war.

BL Cars was custodian of the MG name for eight years between July 1, 1978 and July 6, 1986. Created by chairman Sir Michael Edwardes *qv* as the successor to Leyland Cars, it deliberately downplayed the Leyland name, which he considered should be returned to commercial vehicles from whence it had originated. Edwardes wasted little time in moving MG, in 1978, to the newly established Jaguar Rover Triumph division, where it should have been located in British Leyland Motor Corporation days.

September 10, 1979, forever known amongst MG enthusiasts as **Black Monday,** was the day on which BL Cars announced it had decided to close MG's Abingdon factory. It produced a worldwide outcry and a 'Save Abingdon' campaign was launched by the MG Owners' Club *qv*. Just 20 days later, on Sunday, September 30, MG enthusiasts and owners with their cars from all over Britain descended on London to protest. A 12,246-signature petition was handed in at BL's London office, Nuffield House, Piccadilly, by Cecil Kimber's daughter Jean Cook *qv*.

The Abingdon-based **BMC Competitions Department** sprang from MG's revived racing department, was operational from December 1, 1954, and housed in the old development shop in A Block before moving to the B8 hangar in mid-1955.

Initiated by John Thornley *qv*, it came into being following a memorandum, dated December 13, 1954, by George Harriman *qv* sanctioning the creation of a corporate competition programme. Its first manager was Marcus Chambers *qv*, already at Abingdon as MG's competition manager, with staff supplied by Alec Hounslow's *qv* development department. On January 4, 1955, Harriman nominated Chambers to the committee, at which point he felt that the grouping could be called the British Motor Corporation Competitions Department. The first event it entered was the Monte Carlo Rally, in which an MG Magnette and a TF 1500 kept company with an Austin A90 Westminster. But the best placing was 68th, and that was the Austin!

Later in 1955 a trio of EX 182s *qv* were prepared for Le Mans, but BMC's ban on racing, following the ill-fated TT race in September, saw the emphasis switched to rallying. Experienced rally driver John Gott, then a chief superintentent in the Hertfordshire constabulary, had been appointed captain of the BMC rally team, and Nancy Mitchell, in an MGA and later a Magnette, won the ladies' championship in 1956 and 1957.

Early in 1957, 'Comps', as it was universally known, transferred its operations to the B1 building, where Crusader tanks had been once built. This excised simmering discord between the newly recruited Chambers and Hounslow *qv* of the old regime. Doug Watts was appointed foreman and Tommy Wellman, his chargehand, both proving their worth in the years ahead.

Later, from 1958, the Austin-Healey 100-Six, Abingdon-built from that year, and its 3000 derivative, moved centre-stage. This was to be followed by the Mini, at the expense of the less competitive MGs, although they continued to be prepared for Sebring. With the arrival of the MGB, from 1963 Competitions was allowed to prepare cars for Le Mans although, because of the corporate racing policy, they were entered under the name of the driver.

Chambers retired in 1961, to be replaced by Stuart Turner, a former sports editor of *Motoring News* and an accomplished navigator and co-driver. This marked a golden age for the department with, most memorably, the Mini-Cooper S winning the Monte Carlo Rally in 1964, 1965 and 1967.

Turner left in 1967 to join Castrol, and was replaced by Peter Browning, who had come to Abingdon in 1965 as assistant editor of *Safety Fast qv*. Running costs were some £200,000 per annum, but the creation of British Leyland in 1968 saw a corporate withdrawal from rallying. As a result, Competitions closed on October 31, 1970, although this was not quite the end as the following Special Tuning entry reveals.

The Abingdon-located **BMC Special Tuning Department**, initiated by Stuart Turner, was opened in July 1964 under the management of Glynn Evans, although he was replaced the following year by ex-Morris Engines apprentice Basil Wales, who ran it until 1974.

Its formation followed BMC's first Monte Carlo Rally win, the cars for which had been prepared by the Competitions Department *qv*. The two departments were thus complementary, Special Tuning's function being to sell parts that had been tried, tested and homologated in competition.

A year on, the department could report that it had carried out tuning work on some 200 cars and in excess of 4,000 orders had been received from BMC dealers. Many of these related to parts for the MGB, Midget and Austin-Healey Sprite.

Special Tuning continued to flourish throughout the decade, with annual profits in the region of £100,000, but the creation of British Leyland in 1968 meant that the Triumph range was embraced. It survived the 1970 closure of the Competitions Department, was moved to a new location in C Block in 1971, and was relaunched as Leyland ST in 1974.

Because Competitions had shut, when British Leyland decided to resume rallying in 1975 it was Leyland ST that prepared the works TR7 rally cars in 1976. In the following year came Leyland Motorsport which, as BL Motorsport of 1978, campaigned the Triumph TR7 V8, although these activities ceased in 1980.

In the meantime, the renamed BL Special Tuning parts business had continued trading and remained at Abingdon for eight months after car production ceased. It departed on June 13, 1981, for Cowley and closed down in 1984.

But in 1997 Special Tuning was revived by British Motor Heritage *qv* in conjunction with *Mini World* magazine, and a new catalogue has since been issued that contains more than 1,000 parts. Although inevitably the overwhelming majority relate to the Mini, a notable new product is the MGB front crossmember that was re-engineered for the limited-production MG RV8.

The German **BMW** company acquired The Rover Group on March 18, 1994, and is thus the current owner of the MG marque. It paid £800 million in cash to British Aerospace for the business. Under its auspices MG has been reborn with the arrival of the acclaimed MG*F* of 1995, and surely it can only be a matter of time before MG returns to its traditional American market?

The **Boundary House,** Oxford Road, Abingdon, was the home of Cecil Kimber, his wife and family between 1933 and 1938. It has been a public house since 1962, when it was bought by Morland, the Abingdon brewery, and has since been sensitively extended. Its MG connection is indicated by a bronze plaque on an exterior wall, unveiled in 1990 and initiated by the ever active New England MG T Register.

Once Cecil Kimber's Abingdon home, the Boundary House, now a pub, plays host to MGA, TF and MGB

British Aerospace bought the effectively nationalized Rover Group *qv* that included MG for £150 million on August 12, 1988. It thus returned to the private sector for the first time since 1975. A condition of the purchase was that BAe would retain the business for a minimum of five years, and it sold Rover to BMW in March 1994, five years and seven months later.

British Leyland Motor Corporation (BLMC), effectively a takeover of British Motor Holdings, of which MG was a part, by the Leyland Motor Corporation, came into being on May 14, 1968, and survived until May 20, 1975.

Initially the marque became the MG Car Division of BMC, but British Leyland's management was dominated by former Triumph executives, a reflection that the Coventry-based make had been the first jewel in Leyland's automotive crown in 1961. So on November 1, 1968, MG was relegated to the passenger car Austin Morris Division, whereas logically it belonged in the Specialist Car Division, along with Jaguar, Rover and Triumph. Crucially, responsibility for British Leyland's new sports car passed to Triumph, which produced the flawed TR7 of 1975, and MG's fate was sealed.

British Leyland Ltd was the name given to the business that hitherto had been the British Leyland Motor Corporation *qv*. Effectively a nationalized company, registered on May 20, 1975, it survived until July 1, 1978, when it was renamed BL Ltd. Alex Park had the unenviable task of implementing the government's unworkable Ryder Report that required a centralized corporate structure. As part of the new Leyland Cars, and with marque names downplayed, MG became the Leyland Assembly Plant, Abingdon.

British Leyland Motors Inc, established in 1968, was effectively a merger of Austin-MG, Jaguar and Rover-Triumph interests in America. Located at Leonora, New Jersey, the subsidiary's president was Graham Whitehead, who had served a Wolseley apprenticeship in 1945, moved to America in 1959 and later headed British Motor Holdings (USA) Inc. With the demise of Triumph and MG in 1981, this left it responsible only for Jaguar and in 1983 Jaguar Cars Inc (USA) was established and run by Whitehead until his retirement in 1990.

British Motor Car Distributors, of San Francisco, USA, was formed in 1947 by Kjell Qvale, and in its day was the country's largest British car importer. Qvale was running a Willys Jeep business after the Second World War, when he saw an MG TC in a New Orleans dealership. He decided to dispense with the Jeep franchise and establish an MG one. His first TC was delivered in 1947 and he went on to race its TD successor. As sales increased, Qvale began to diversify into Austin, Morris, Jaguar, Rolls-Royce and Austin-Healey. It was the withdrawal of Austin-Healey from the American market that led him, in 1970, to acquire 84 per cent of Jensen *qv* to produce the Jensen-Healey sports car of 1972, which ceased production in 1976, taking Jensen with it.

The **British Motor Corporation** (BMC) was custodian of the MG name from its inception on February 25, 1952 until 1968. Ostensibly a merger, but in reality a takeover of Morris Motors by the rival Austin Motor Company, BMC was Britain's largest indigenous car maker and was also responsible for the Austin-Healey, Riley and Wolseley makes.

Its chairman until 1961 was former managing director of Morris Motors, Leonard Lord *qv*, who was succeeded by George Harriman *qv*. A merger with Jaguar on December 14, 1966, resulted in the creation of British Motor Holdings (BMH), which in its turn was taken over by the Leyland Motor Corporation to form the British Leyland Motor Corporation *qv* in 1968.

Established in 1985, **British Motor Heritage** is a Rover Group *qv* subsidiary and best known for the remanufacture of the MGB and Midget bodyshells. Created as the cash-generating arm of the British

British Motor Heritage's manufacturing manager, David Bloomfield, with an MG RV8 bodyshell at Faringdon.

Motor Industry Heritage Trust, it manages the Association of Heritage Approved Supplies parts scheme initiated by the Trust in 1980. At the instigation of David Bishop *qv*, it decided to put the MGB roadster body back into production. Premises were established at Faringdon, Oxfordshire, conveniently located between Rover Pressings at Swindon and BMH's Cowley office. Introduced in 1988, the right-hand-drive roadster's chrome-bumpered hull was followed in 1990 by the MGB GT shell and the pre-1974 American-specification roadster. The Midget, and thus Sprite shell, and right-hand-drive 'rubber'-bumpered B roadster body appeared in 1991. Subsequently a move was made to larger premises at Range Road, Cotswold Business Park, Witney, Oxfordshire, from where BMH currently operates.

Roy Brocklehurst (1932-1988) succeeded Syd Enever as MG's chief engineer in 1971, and spent 26 years with the company working on the TD, MGA, MGB, MGC and MGB GT V8 models.

Brocklehurst - determined, able and humorous - always said that he'd joined MG by mistake. He grew up in the nearby town of Didcot, and when he left school at the age of 15, in 1947, he cycled over to the nearby Atomic Energy Research Establishment at Harwell to inquire whether it was offering any engineering apprenticeships. It wasn't, so he carried on until he reached MG and, in June, began work there as a design apprentice. He was soon drawing the special three-throw crankshaft for Goldie Gardner's 1949 500cc record attempt in EX 135.

Brocklehurst obtained his theoretical training at the Oxford College of Technology prior to doing his National Service between 1952 and 1954. The latter year was when the design office was reopened at Abingdon and, in June, Roy was made a layout draughtsman. He found that he had some new colleagues, Enever having recruited four draughtsmen, two chassis and two body men, from Cowley.

He was promoted to the post of chief draughtsman in September 1956, by which time he had become deeply involved in the layout of the MGB's mechanicals. Brocklehurst initially designed the B with a coil-sprung rear axle rear suspension, but this was dispensed with when handling proved unpredictable, and conventional leaf springs, in the manner of the MGA, were fitted.

He relished his work on the B, was made project engineer in 1961, and in December 1964 assistant chief engineer, so becoming Syd Enever's heir apparent. He was also responsible for the MGC's unique torsion-bar independent front suspension. Later he robustly described the model to me as "a political nonsense, an understeering pig".

After he took over from Syd Enever as chief engineer in April 1971, Roy was also responsible, at the instigation of British Leyland's technical director Harry Webster, for overseeing the creation of the first experimental MGB GT V8 conversion, although the car was seen into production by his successor, Don Hayter *qv*.

By this time Brocklehurst had moved to Longbridge, having been promoted, in July 1973, to the post of Austin Morris' chief engineer, vehicle engineering. As such he became closely involved in the design of the Austins Metro, Maestro and Montego. In 1980 he became chief engineer, advanced vehicles of Gaydon-based BL Technology and retired when that operation was wound down in 1987. He was looking forward to retirement at his Hayling Island holiday home when he died suddenly following a stroke at the age of 55.

C

BMC's six-cylinder **C Series** *qv* pushrod engine of 1954, designed and produced by Morris' Engines Branch *qv*, was not used by MG, but powered the Austin-Healey 100-Six of 1956 and its 3000 derivative, Abingdon-built from 1958. The C did, however, bequeath its ultimate 83 x 89mm dimensions and 2,912cc capacity to the 29G unit used in the MGC *qv*.

The 746cc **C-type,** known in its day as the Montlhery Mark II MG Midget, was produced to commemorate George Eyston's successful 101mph record-breaking run in EX 120 *qv* at the French circuit in September 1931. Available in both Powerplus supercharged and unblown forms, it was manufactured between April 1931 and June 1932 and its chassis numbers ran from C.0251 to C.0294. A total of 44 were built.

Carbodies of Coventry was the principal supplier of MG bodies between 1924 and 1935. Car Bodies, the original rendering, was founded in 1919 by Robert (Bobbie) Jones against a background of 30 years' experience of coachbuilding. Born in Bury, Lancashire, in 1873, he studied engineering at Manchester Technical College, and by coincidence this was where Cecil Kimber *qv* later completed his education. But Jones had to walk nine miles each way.

By the age of 16, in 1889, he was apprenticed to a body maker, but 1904 saw him at Beeston, Nottinghamshire, working on the bodies of the first Humber saloons. In 1907 he moved to Coventry to become works manager of Charlesworth Bodies *qv*, transferred to Hollick and Pratt *qv* in 1912, but in 1913 went north again to Watsons, the Liverpool coachbuilders.

Prior to his establishment of Carbodies in 1919, Jones became coachbuilding manager for Mann Egerton in Norwich. He returned to Coventry to take over the coachbuilding business of Gooderham and Sons, of Foleshill, on a small site in Old Church Road. Its first commission from Cecil Kimber was The Morris Garages Chummy *qv* of 1922, and this led to special bodies for the MG Super Sports 14/28 and 14/40. A handful of the Garages' employees were thus employed to drive the 50 or so miles to Carbodies' by then larger premises in the Holyhead Road, near the Alvis factory.

But as demand for the M-type increased, so necessitating MG's move to Abingdon, the two-seater fabric bodies, for which Carbodies charged MG £6.10s (£6.50) apiece, were delivered by lorry packed in crates, each of which contained three bodies. With Morris Motors' takeover of MG in 1935, Kimber had to sever his links with Carbodies, and from thereafter bodies were supplied by Morris' Bodies Branch *qv*, also located in the city.

Whilst Carbodies had been geared to the volume production of MG bodies, it also supplied coachwork to a variety of the city's other car makers, namely nearby Alvis, Hillman and Humber, as well as for Morris' short-lived Ten Six Special, initiated by Len Lord *qv*.

During the war a wide range of vehicles were bodied, including Field Marshal Montgomery's caravan. In 1948 Carbodies produced the prototype Austin FX3 taxi, and this was gradually to become its mainstream product. In 1955 the business was bought by BSA, so securing Daimler's body supply, and it also produced Ford's low-volume convertibles.

When Jaguar acquired Daimler in 1960, its BSA parent retained Carbodies, but on the 1973 collapse of motorcycle manufacture, its non-motorcycle interests were bought by Manganese Bronze, which has owned it ever since. Now renamed London Taxis International, it currently produces some 2,000 a year, making this one-time MG supplier one of the most productive of British-owned car makers!

A building which is now named Larkhill House, in **Cemetery Road,** Abingdon, used to be MG's offices from 1929 to 1980 and is the only

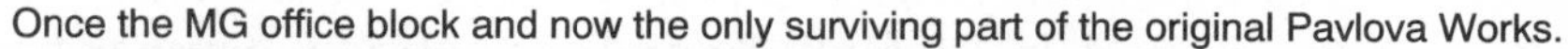

Once the MG office block and now the only surviving part of the original Pavlova Works.

surviving remnant of the original Pavlova Works *qv*. What became MG's A Block was demolished in 1997 and the leather factory followed in 1998. The refurbished building has been occupied since 1980 by the construction and facilities management company Barwick, its owner since 1994.

Marcus Mordaunt Bertrand Chambers (born 1910), was from 1955 until 1961 manager of the Abingdon-based BMC Competitions Department *qv*. He brought with him a wide experience of racing and the motor trade. Educated at Stowe and in France, in 1928 Chambers joined ICI, but between 1931 and 1938 he variously worked for Lex Garages, the Larkhill Estate Garage, and Terence Windrum and Garstik.

In 1938 he opened his garage business, Marcus Chambers Motors, at Chichester, Sussex. An active motor sport competitor since 1931, he participated in races, hillclimbs and speed trials and drove HRGs at Le Mans. He was placed 10th in 1938, and the following year attained a creditable 13th position and first in class.

After war service in the RNVR, Chambers became manager of North Downs Engineering Company and worked briefly in HRG's racing department. In 1949 he moved to Tanganyika, Africa, with the Overseas Food Corporation, better known as the controversial ground nuts scheme, and then with the Colonial Development Corporation in the West Indies.

In the summer of 1954 Chambers returned to Britain and, following a meeting with John Thornley *qv*, was later that year appointed MG's competition manager, just prior to the formation of the BMC Competitions Department. He took up that appointment in March 1955.

Marcus Chambers held the position for seven years, but left in 1961 and returned to the motor trade as service manager for Appleyard of Bradford. Subsequently, in 1964-69, he was competition manager for the Rootes Group. His book, *Seven Year Twitch*, recounting his years at Abingdon, was published in 1962, then was extensively revised, expanded and republished by Motor Racing Publications under the title *Works Wonders* in 1995.

Hubert Noel Charles (1893-1982) was the talented graduate engineer who had overall responsibility for the design of all MG road and racing cars produced between 1930 and 1938. Born in Barnet, Hertfordshire, Charles was educated at Highgate and obtained a BSc degree in engineering from London University. A mathematician of formidable ability, he listed the subject as one of his hobbies.

In 1914 Charles worked on a fuel-injected Triumph motorcycle for a Brooklands record attempt, but with the outbreak of the First World War joined the Royal Naval Air Service as a mechanic. Later he switched to the Royal Air Force, where he attained the rank of Captain, and was twice mentioned in despatches.

With the coming of peace in 1919, Hubert Charles was employed in the technical sales department of Zenith carburettors until 1920, and the following year moved to the then London-based Automotive Products, where he worked in a similar capacity. He moved to Morris Motors at Cowley in 1924, where he was employed as a technical assistant in the production department.

It was when there in 1925 that he met Cecil Kimber *qv*, who immediately recognized his potential, and Charles variously worked on the Oxford-built 14/28, 14/40, M-type and 18/80 models. This was undertaken mostly in the evenings and at weekends, often at Kimber's Oxford home at 339 Woodstock Road.

Following the company's move to Abingdon, in 1930 Charles was appointed MG's chief draughtsman, in truth its chief engineer, and ran the drawing office there until it closed in 1935. His proteges included Syd Enever *qv*, Jack Daniels *qv*, and Adrian Squire *qv*.

As John Thornley has so perceptively observed in *Safety Fast:* "In this period more than a dozen distinct models were designed and produced, many of them with numerous derivatives. And this was

Hubert Charles as photographed by Wilson McComb in 1965.

achieved by Charles working with the merest handful of devoted designers and draughtsmen. To achieve so much, so much that was innovative, so much that was *right* - surely here was the touch of genius."

This was displayed in his last MG racer, the single-seater R-type of 1935 that sparkled with ingenuity from its light but immensely strong Y-shaped chassis to its all-independent torsion-bar suspension. It clearly required some refinement, rolling excessively on corners, but almost immediately Charles had to return to Cowley.

There he oversaw the new generation of pushrod-engined MGs, namely the SA, TA and B and VA models. He was also responsible for the ingenious front suspension of the new unitary construction Morris Ten of 1938 that employed flexible leaf springs in conjunction with a high-mounted anti-roll bar.

It was at this time that Charles, along with A V Oak, undertook a series of experiments on body structures and their relative stiffness in relation to suspension and handling. This was the work on which Alec Issigonis *qv* was to draw when he arrived at Cowley in 1936, and was later able to incorporate in the Morris Minor of 1948.

Charles' thinking is also reflected in that remarkable vehicle in its use of longitudinal torsion bars, a layout he had employed on the MG R-type racer. As he later put it, Issigonis was "very receptive and eventually fell for my own great interest in torsion bars and their associated general effects". The two engineers did not get on!

But after three years at Cowley, Charles returned to the aircraft industry in 1938 to become chief engineer for Rotal Aircrews, but in 1941 he joined Austin as a development engineer. There he was responsible for the coil-and-wishbone independent front suspension system for a new generation of postwar Austins. Paradoxically, this pioneer of torsion bars had become wedded to coils springs. And his inspiration was none other than Alec Issigonis, who had advocated them prior to his conversion to Charles' torsion bars! "I fell for his studies of coil springs and all their associated effects."

Whilst at Longbridge, Charles worked as a development engineer on the 1.2-litre engine used in the 1947 Austin A40 which in its later B Series *qv* form powered the MGA and MGB. Sadly, this supremely talented engineer clashed with Austin's chairman, Leonard Lord *qv*, and left in 1946.

Becoming a consultant mechanical engineer, in this capacity he was retained by Cam Gears and Norton Motorcycles, where he and Joe Craig worked on a type of carburettor control by means of a hot wire analyzer of its exhaust gases. In 1953 Charles became a partner in an experimental engineering business, Manley and Charles, at Eynsham.

In McComb's *qv* words: "Original thinking was characteristic of his work, which often went unappreciated because he was ahead of his time." It was a view shared by that shrewd observer of the motor industry, *Autocar's* Ronald Barker, who regarded Charles as an "extraordinary genius". He died in Oxford on January 18, 1982, at the age of 88, and his reputation can only grow in stature with the passing years.

Charlesworth Motor Bodies, which produced factory-approved MG bodies in 1936-39, was established in Much Park Street, Coventry, in 1907. Its works manager was Robert Jones, who went on to establish Carbodies *qv*. Later, in 1931, the firm was a victim of the recession and was reformed as Charlesworth Bodies.

Charles Reynolds became managing director and a major shareholder, and Charlesworth enjoyed a revival with an impressive range of coachwork, perhaps the work of A J Cannell, an associate of Cecil Kimber *qv*. Its work for MG began with a batch of handsome four-door tourers; some 90 were produced, in the Alvis idiom on the SA chassis and for the later WA. A two-door tourer was offered on the VA frame. When Kimber lost his job at MG in 1941, Reynolds offered him employment to reorganize a Gloucester-based aircraft factory for which he was responsible. He remained there until the end of 1942.

After the war Charlesworth became briefly involved with building the coachwork for the ill-fated Invicta Black Prince saloon of 1946, but only some 25 were produced and in 1948 the firm was bought by Lea-Francis and so survived until its demise in 1954.

Whilst some MGs were exported in an unassembled state in the 1930s to Australia *qv* and Eire, after the war Booth Bros of Dublin assembled 84 TCs exported in **CKD** (for Completely Knocked Down) form and repeated the exercise with the TD (98 cars). British Car Distributors in South Africa also took 345 TDs, whilst Automoviles Ingleses in Mexico assembled 75 left-hand-drive TFs.

With the arrival of the higher-volume MGA, kits were sent to Australia, South Africa, Eire, Mexico and to J Molenaar of Amersfoot, in The Netherlands, and it was the same story with the MGB. The largest market for such kits was Australia, and between 1962 and 1972, when the operation ceased, a total of 9,090 MGBs had been assembled by BMC (Australia) at its Victoria Park, Sydney works. A total of 104 examples were similarly produced at BMC's plant at Malines, Belgium, between 1964 and 1968. Eire continued to take unassembled MGBs between 1966 and 1971, although this amounted to just 404 cars.

On July 22, 1926 H N Charles *qv* set down his ideas in a letter to Cecil Kimber *qv* for what he called a **Comparator.** Designed to test cars statically, it resembled a rolling-road, was first used at Bainton Road, subsequently at Edmund Road and then at Abingdon.

Harold Connolly was not only the illustrator, from 1929 until 1939, of MG catalogues, more significantly he collaborated with Cecil Kimber *qv* on MG's styling. Connolly came to the latter's attention for the work he undertook on Morris' catalogues and took over the MG commission from Leslie Grimes. He also illustrated the company's leaflets and advertisements.

M-type: a typical rendering by Harold Connolly.

He later recalled Kimber saying he liked his work because "the cars looked as if they were made of metal – the air-brush drawings of cars in those days made them look like silk stockings . . . Kimber always said a sports car should look fast even when it's standing still."*

Connolly's cars are charmingly peopled by fragrant young girls, no doubt fresh from a strenuous game of tennis, who are often accompanied by sports jacket-clad beaus. He was paid five guineas (£5.25) apiece for his work that delights to this day.
* *MG by McComb*

Cecil Kimber's younger daughter, **Jean Cook** (born 1925), is a doughty champion of her father's memory. She wrote the foreword to Wilson McComb's magnificent MG history, published in 1978, later essayed a fascinating profile for *The Kimber Centenary Book* (1988) and more recently, as more of the story unfolds, for *The MG Log* (1993). She has also written many magazine articles on her famous father. Fittingly she handed in the petition to save MG and the Abingdon factory when the MG clubs gathered in London in 1979 to protest at the impending closure of the Abingdon factory.

'The Corsica' was Cecil Kimber's own MG, a 1934 K1 Magnette tourer with a 1.3 KD-type supercharged engine and black drophead four-seater coachwork by Corsica, hence its name. Despite being capable of 100mph, the car's engine fumed badly, a problem experienced by Michael Rabone, who bought the car from Kimber in 1936. It was returned to the factory for this reason on several occasions, although service manager John Thornley *qv* diplomatically countered that it was "essentially a racing chassis". Rabone had the coachwork mildly updated by Corsica in 1937 and the car happily survives, although in unsupercharged form.

Ken Costello was responsible for the concept of uniting Rover's V8 engine with the MGB GT, so providing the inspiration for the MGB GT V8 of 1973. An enthusiastic racer, Costello, from Farnborough, Kent, was seen in the 1960s at the wheel of a Mini and he was sufficiently successful to find a small but steady market for similarly enhanced Mini racers.

But in 1969 his life changed when he first encountered the General Motors-designed alloy V8 *qv* on the premises of Piper Cams, at nearby Farningham. As he later told *MG World:* "I was able to lift it on my own, and immediately wondered if it would go into the MGB." The first conversion, in which the Oldsmobile version of the V8 was shoehorned into a red MGB roadster, was built between June and November of 1969.

Ken undertook a second conversion in a newer green MGB GT, although this used the Anglicized Rover unit. It fitted remarkable well after modifications, which included changes to the steering as the original fouled the new fabricated exhaust manifolds, and the fitment of a 3.07:1 rear axle. A 9.5in MGC clutch also featured. The most obvious external change was the use of a glassfibre bonnet to clear the engine's obtrusive SU carburettors.

This car was effectively the prototype Costello MGB V8 and, from thereon until 1973, and even into the 1990s, a steady trickle of cars emanated from the workshops of Costello Motor Engineering, in Farnborough Way. Not only was the Costello V8 some 90lb lighter than the production B, in 1972 *Autocar* recorded a top speed of 128mph in a car that cost £2,616, which compared with £1,459 for a new GT. The professionally executed work was undertaken on both MGB GTs and some roadsters, which invariably were fitted with a black-painted eggbox grille. At the rear, a *V Eight Costello* badge was mounted above the existing lettering.

Inevitably the cars attracted the attention of British Leyland, which led to it commissioning a car for assessment and Costello duly prepared a left-hand-drive GT finished in Harvest Gold. At Leyland's instigation, Abingdon responded by producing its own prototype, which paved the way for its own MGB GT V8 *qv* of 1973. Unfortunately, the appearance of the factory car damaged Costello's business, a problem that was compounded by British Leyland's refusal to supply him with complete Rover engines. It informed dealers that they were required to take the old unit in part-exchange.

In the meantime, Costello had effected modifications to the original design, and the post-1973 cars have been accorded 'Mark II' status and are thus much rarer than the originals. These usually had used Buick or Oldsmobile blocks, but no matter. With some 200 cars completed, Ken Costello's place in MG history is now assured.

Jacques Coune, of Brussels, was responsible for the Berlinette MGB 1800 produced between 1964 and 1966. Introduced prior to the appearance of the MGB GT in 1965, it was a professional Ferrari-inspired roadster conversion which retained the metalwork to the rear of the car, with the back of the body and roof made of glassfibre.

The work was undertaken at Coune's premises at 286-290 Avenue de la Couronne, Brussels, and the prototype red steel-bodied car was unveiled at the capital's 1964 motor show. After three cars had been made, lighter glassfibre was adopted, with Coune mostly buying complete B's from MG's Belgium importer Molenaar, although some customers' cars were converted. The higher windscreen and rear window were courtesy of the Renault 8 and the Berlinette sold for the equivalent of £1,285. By the middle of the year, the firm claimed to be producing 12 to 15 examples a month.

According to Coune, all were left-hand-drive Bs with the exception of a single British delivery, ordered by Walter Oldfield, managing director of BMC's Nuffield Press. On completion, this right-hand-drive car was tested by factory personnel and by Sir George Harriman *qv* and Alec Issigonis *qv*, with a view of putting it into production but the concept was rejected for being 'too Italian' in appearance.

Unfortunately for Jacques Coune, MG introduced its own GT at the 1965 London Motor Show, which immediately negated the appeal of the Belgian version, not least because it retailed for £660 more. So production ceased after about 56 Berlinettes had been built.

This was not before Coune had tried again with a one-off

luxuriously equipped MGB-based Targa version of the coupe that appeared at the 1966 Brussels motor show. But by 1970 Coune had left the motor trade and in the 1980s he became the co-founder of a Brussels motor museum.

Henry Edward Cecil ('Cec') Cousins (1902-1976), who always reckoned that he was MG's first employee, became works manager at Abingdon in 1944, a position he held until his retirement in 1968. Oxford-born 'Cec', as he was universally known – lanky, 6ft 4in in his socks – was educated at St Paul's School in the city, and at the age of 14 was apprenticed with a local business, T G West and Sons, who were steam and agricultural engineers.

The four years Cousins spent with Tommy West provided an excellent grounding in all mechanical matters; the business even produced the Atalanta, its own motorcycle. After three and half years Cousins left and, on January 5, 1920, he joined The Morris Garages' premises in stables at The Clarendon Hotel, Oxford, as a motorcycle 'improver'.

Late in 1922 he moved to the business' Longwall garage *qv*, where Cecil Kimber *qv* was beginning to produce the Morris Garages Chummy *qv* and Cousins followed him in 1923 to the small Alfred Lane mews garage *qv* to continue production. He thus became, in his own words, "responsible for all production and service activities while it remained under the control of Morris Garages".

With the move to Abingdon he initially ran the experimental department, but soon became works superintendent and handed over his first post to Reg Jackson *qv*. As John Thornley *qv* remembered in *Safety Fast:* "What, in fact, he did was to act as a contact man between the designers under H N Charles and the prototype shop, or as this haven of glamour and witchcraft was known, 'experi'." He was also able to become involved in racing and was regularly pictured with MGs on the track.

During the war Cousins became aircraft superintendent and was appointed works manager in 1944. It was in that capacity that he led the group that seized the initiative and created the TD when sales of the TC began to flag. He was to repeat this exercise with the TF.

'Cec' Cousins retired in May 1967, the year prior to the takeover of BMC by British Leyland. In 1966 he had acted as general manager during John Thornley's illness.

Let the last word on 'Cec' go to Thornley: "His preoccupation was the welfare of folk on the shop floor and the fact that he was always straight as a die were the two qualities which contributed most to the respect shown to his opinions, and woe betide anyone who uttered, in his hearing, things critical of MG which he thought to be unjust."

The Oxford district of **Cowley** has intermittent associations with MG from the make's 1923 origins until the low-volume MG RV8 *qv* ceased production in 1995. Morris began to manufacture cars there in 1913 and it was thus the source of MGs produced between 1924 and 1929, cars that were essentially modifications of production Morrises. In the years 1927-29 manufacture was based in a purpose-built plant at Edmund Road *qv* at Cowley, near the factory.

Following the discontinuation of racing, MG design was transferred in 1935 to Cowley and it continued to be so responsible until 1954, when the Abingdon drawing office was reopened. Models such as the SA, VA, WA, Y-type and Z Series Magnette saloons and TA and TB sports cars were designed there.

Cowley also assembled the MG-badged saloons of the BMC era, namely the low-volume Pinin Farina-styled Magnette of 1959-68 and the popular MG 1100/1300 of 1962-73. It also produced 476 examples of the Mark III Midget between January and March 1967.

After the closure of the Abingdon factory, the MG versions of the Maestro and Montego were Cowley-built. From 1990 the factory was radically altered in a two-year £200 million redevelopment programme in which the original South Works was demolished. Car production was then concentrated on the south side of the Oxford Eastern Bypass, which had previously housed Pressed Steel *qv*.

However, MG sports car production was Cowley-based between 1992 and 1995, when the MG RV8 was produced there in a special low-build facility established in a former body-in-white panel store. Currently the plant is Rover's Large Car manufacturing facility and so the MG*F* is built at Longbridge *qv*.

Edmund Road, **Cowley,** was MG's home for two years, from September 1927 until September 1929. As MG was to proclaim, it was "unique in being the only factory in the world entirely devoted to the production of Sports Cars".

As the Bainton Road *qv* radiator factory was clearly too small for his needs, Kimber later recalled that he went to see William Morris *qv*, who quickly grasped the problem and asked him what he intended to do. "I replied 'I propose to spend £10,000 and build another factory'. He said 'Where do you propose to build it?' I told him, and he said 'Right, go ahead.' He did not see a plan or specific action and we actually spent £16,000, and all out of the profits we had made."

The Morris Garages' works was built by Kingerlee and Company. Carl Kingerlee was a friend of Morris' and his secretary in his later years. The task took six months to complete and the factory was conveniently close to the Cowley plant. Initially the work was concentrated on modifying Morris chassis and engines and it was only

MG's first purpose-designed factory at Edmund Road, Cowley.

with the arrival of the 18/80 in 1928 that a rudimentary drawing office was established.

The last of the 14/40s, the first of the 18/80s and the M-types were produced at Edmund Road in chassis form and, as before, there was no body shop. Having been road-tested with a ballast body to represent the weight of the coachwork, the chassis were fitted with trade plates and then driven to Carbodies *qv* in Coventry to receive their coachwork. The cars were sprayed at a works down the Cowley Road in Leopold Street, Oxford. It was a building previously occupied by a bus company that used to keep its open-topped vehicles there.

In addition to the creation of MG's first production line, a service department was established at Edmund Road. Previously this had been undertaken at the Morris Garages' Longwall, Clarendon and Merton Street depots.

After MG's departure for Abingdon in September 1929, the Edmund Road works was used for a variety of purposes including, just prior to the outbreak of war, an expanding Radiators Branch. It was demolished in the mid-1970s and the site is now occupied by a housing estate.

The famous **Cream Crackers** helped to keep the MG name alive in the public eye by successful participation in cross-country sporting trials between 1935 and 1938. Factory-sponsored to varying degrees and so named because of their distinctive chocolate and brown bodywork, initially a trio of alloy-bodied PBs, registered JB 7521, JB 7524 and JB 7525, were used. For 1937 a team of factory-prepared and mildly tuned TAs, ABL 960, ABL 962 and ABL 964, were produced and fitted with Wolseley gearboxes. To counter resentment, the cars were sold to their drivers and repurchased at the end of the season.

In 1938 came a new batch of more heavily modified TAs, BBL 78, BBL 79 and BBL 80. It proved to be a good season and for the second year running the Crackers won the MCC Team Championship.

Frederick Gordon T Crosby (c1885-1943), was arguably the greatest of all motoring artists, whose work enhanced the pages of *The Autocar* from pre-1914 days throughout the inter-war years. The prolific Crosby also undertook many assignments on a freelance basis, which allowed him to live comfortably in the fashionable district of Sunninghill, Surrey, and he enjoyed a round of golf at nearby Wentworth.

Well-known throughout the British motor industry, 'Gorby', as his intimates knew him, was a friend of Cecil Kimber *qv* and owned a 1930 18/80 de luxe saloon WL 9253 that survives in America. He also designed the MG Car Club's *qv* badge.

But it is the front cover work he undertook for *The Autocar* to accompany MG advertisements that is particularly impressive, most notably that of an 18/80 saloon and an MG Midget (January 24, 1930) and the issue of May 2 the same year, which shows the 18/100 on the Brooklands banking. There were at least two renderings of this scene and in the second version that he undertook for the firm the big MG is less threatened by an approaching car! It graced the sitting room of John Thornley's Abingdon home for many years.

Perhaps the most memorable example of Crosby's MG work was *The Autocar's* front cover of April 29, 1932 that featured the understated but alluring MG Girl. Interestingly, Crosby was less happy with his renderings of the human figure, so the face was the work of his son Peter, whilst the model was Crosby's wife Marjorie. Happily, the original artwork survived and was sold at auction in 1990 for £10,500.

Peter M G Crosby, who showed himself to be as accomplished an artist as his father, also undertook work for MG, most notably illustrating 'About George', a promotional cartoon strip. He served in the RAF during the war, and his death when a Flying Officer in 1943 resulted in his already depressed father taking his own life in August of that year at the age of 58.

The Crown and Thistle hotel in Bridge Street, Abingdon, was used as a background for many MG publicity photographs. This early 17th century coaching inn, so named to represent the union of England with Scotland, was frequented by MG personnel from Cecil Kimber onwards. As such it became known at the works as the Crown, Wheel

The Crown and Thistle hotel, Abingdon, used as the background for many an MG publicity photograph and one-time MG factory entertaining venue and watering hole!

and Pinion . . . Part-owned by Morland from the late 19th century, it closed in 1973, was reopened as a Berni Inn the following year and is now a Scottish and Newcastle hotel. I can recommend the English breakfast!

In the 1930s, a **cup,** made entirely of MG components, was awarded annually to the make's most successful sales representative. Designed at Abingdon, its body was made of an inverted headlamp and bore an MG radiator badge with the handles provided by two bent connecting rods. It was mounted on a base made of a crownwheel and the winner's name was inscribed on a succession of plinth-mounted octagonal shields. Where is it now?

Alan G Curtis (born 1926), one-time chairman of Aston Martin Lagonda, was responsible for initiating a consortium to save MG following news of the factory's closure in September 1979. Announced in October 1979, it included British Car Auctions and civil engineers Norwest Holst. On March 31, 1980 came an announcement that it had reached agreement with BL, but as Curtis later told *Autocar,* "things started to go wrong" in May and the worsening recession provided the knockout blow. Curtis confirmed that John Symons, former head of BL's Pressed Steel Fisher and chief executive of Aston Martin, would have headed the team of the new MG production company.

D

The underpowered and rather inelegant four-seater version of the M, the 847cc **D-type** was produced between October 1931 and June 1932. Total production amounted to 250 cars, of which 208 were four-seaters, 37 Salonettes and five bodied by other coachbuilders. Chassis numbers ran from D.0251 to D.0500.

George Eyston's *qv* **Dancing Daughters** team of all-lady drivers, who drove a trio of P-types at Le Mans in 1935, was so named by a journalist after a stage act of that name.

Jack Daniels (born 1911), who joined MG at its new Edmund Road factory *qv* in 1927, was the company's first unindentured apprentice. Later he worked closely with Alec Issigonis *qv* and had the reputation for being an outstanding draughtsman. One of his abiding memories was the sight of Morris chassis "covering the half-mile from the Cowley factory under their own power, two pullling five in a line". His best subjects at the Oxford Central school had been woodworking and engineering drawing, and after about two years, Keith Smith, who ran the MG drawing office, asked him to help out. Facilities were primitive and they had no printing equipment to reproduce their completed drawings, so they had to use the services of an architect in Headington.

Daniels, and Smith's subsequent replacement, George Gibson, were amongst the first on the scene at the Abingdon drawing office. Much influenced by H N Charles *qv*, "he was my real tutor and a very sound engineer," Daniels moved to Cowley when the racing programme ended in 1935. From 1937 he worked with Alec Issigonis by transferring the latter's freehand sketches into full working drawings. Involved in the conception of both the Morris Minor and Mini, Daniels delighted in quipping that the designs "were his inspiration and my perspiration!" His last assignment for what had become BL Cars was on the LM 10 project that emerged in 1983 as the Austin Maestro hatchback.

Sydney Charles Houghton 'Sammy' Davis (1887-1981), the popular sports editor of *The Autocar*, was manager of the Eyston Dancing Daughters *qv* team of P-type MGs that ran at Le Mans in 1935. Educated at Westminister School and University College, London, Davis served an apprenticeship at Daimler at Coventry with Frederick Gordon Crosby *qv*, but left in 1912 to join publisher William Iliffe, based in the city, for the launch of *Automobile Engineer* and soon found himself also working for its *Autocar* stablemate.

After war service in the Royal Naval Air Service and RAF, Davis rejoined *The Autocar* in 1919 to become sports editor and wrote under the *nom de plume* of *Casque*, which is French for helmet. He had raced motorcycles before the war and cars from 1921, his most celebrated victory being in 1927, when he co-drove the 'Old Number 1' Bentley that won Le Mans.

Davis' celebrated road test of a J2 in which the car exceeded 80mph was the result of special tuning at Abingdon! An artist of note, he also initiated and designed the prized silver-winged MG overall badge used by drivers George Eyston, Bert Denly, Wally Handley, Norman Black, Eddie Hall and Goldie Gardner. Racing mechanics Alec Hounslow, Henry Stone, Bob Scott, Dump Barrett and Jack Mathews all proudly wore their 'wings'.

Sammy finally retired from *The Autocar* in 1950, but he continued a further indirect contact with MG when he was appointed Austin's adviser on motor sport and continued to do so during the BMC years. When Abingdon reintroduced its racing programme he found that many of the individuals at the factory he had known in the inter-war years were still there in the 1950s.

There was a **deliberate mistake** in the published specification of the KD Magnette of 1933-34 vintage. This was because Abingdon made strenuous attempts to disguise the origins of the 1,271cc six-cylinder engine which was identical to that used on the Wolseley Hornet. In the catalogue Kimber therefore added a millimetre to the engine's 83mm stroke, making it 84mm. This contributed 15cc to the capacity which then read 1,286cc. H N Charles was not amused.

John Frederick Dugdale (born 1914) has had a long involvement with MGs, which began in 1933 when he joined *The Autocar* as assistant to sports editor Sammy Davis *qv*. Born in Weybridge, Surrey, and educated at Rugby, Dugdale was the magazine's assistant editor in the 1938-40 era. From 1933 onwards he was therefore well acquainted with Kimber and in 1935 bought a new N-type Magnette that he raced at Donington, Brooklands and Crystal Palace in 1937-38. In 1938 and 1939 he was a member of the parties that went to

Germany to witness Goldie Gardner's *qv* successful record-breaking runs with EX 135 (q.v).

During the Second World War Dugdale was a Captain in the Royal Army Service Corps and, after hostilities, he moved to America in 1949, where he represented Rootes until 1954. He then became vice-president of Jaguar Cars Inc, but after four years he left to become the North American representative of the Society of Motor Manufacturers and Traders.

In 1966 John Dugdale rejoined Jaguar as vice-president of advertising and public relations. The various mergers within the British motor industry led to him becoming national product publicity manager for British Leyland in the US until he retired in 1980. It was in this capacity that, in 1979, he launched the company-backed *MG Magazine qv*.

John Dugdale is the author of the acclaimed *Great Motor Sport of the Thirties* (Gentry Books, 1977) which, although out of print, is today much sought after by MG enthusiasts.

E

As chairman of BL Cars *qv*, **Sir Michael Edwardes** (born 1930), the diminutive South African lawyer, had the unenviable job of discontinuing the MGB and closing MG's Abingdon factory. Appointed in November 1977 on a three-year secondment, he eventually stayed for five. He inherited the decision that the corporate sports car was to be the Triumph TR7 of 1975, and MG was left to continue assembling the ageing B.

Edwardes' many problems were exacerbated by the election of Margaret Thatcher's Conservative government, which saw the value of the pound soar to the extent that, by the beginning of 1981, it was worth 20 per cent more than it had been on election day. This played havoc with exports; 80 per cent of MGBs were sold in America.

The MG closure was part of the diplomatically titled Recovery Plan, announced in September 1979, which was designed to reduce BL's excess capacity and involved the loss of 25,000 jobs. This included the closure of the Speke 1 factory in Liverpool, the Castle Bromwich pressing facility and the Triumph plant at Canley. But as Edwardes later declared in his autobiography, *Back from the Brink:* "The decision to stop MG car production created more public fuss and misunderstanding than anything in the whole five years - even greater than wholesale factory closures and job losses."

Albert Sydney (Syd) Enever (1906-1993) was much more than MG's chief engineer between the years 1954 and 1971. His abilities ranged over the full gamut of racing cars to record-breakers, production models and even styling. From the time he joined the firm in 1930 until his retirement at the age of 65 in 1971, there were few aspects of the MG Car Company of which he was unaware. To say that he was the archetypal 'fag packet engineer' - he was forever jotting down ideas on scraps of paper - should not be taken as criticism, it was in recognition of his astonishing creativity.

For Syd was a supremely accomplished intuitive thinker and, as John Thornley has said: "It is not correct to describe him as an

Syd Enever (left) in company with that great MG driver, Captain George Eyston.

untrained engineer because he trained himself: he had an insatiable curiosity about why things happened, a curiosity that went beyond purely engineering and automotive matters."

Syd Enever was born at Colden Common, Hampshire, one of eight children. His father was an ironmonger's assistant and was later apprenticed as a maker of stained glass windows. But Syd's parents separated when he was three, and the family later moved to Oxford, where his mother, Maud, opened a theatrical boarding house near Oxford Theatre.

He displayed a precocious mechanical aptitude and, on leaving school at the age of 14 in 1920, Syd's headmaster, a Mr Benson, recognized it, to his eternal credit, and found him a job in The Morris Garages *qv* salesroom in Queen Street at 12s 6d (62p) a week. There young Enever served as an errand boy, parts courier and parts polisher. After a probationary period of a year, he went to work at one of the garages behind the Clarendon Hotel.

At the age of 20, in 1926, Syd acquired a 3.5hp BSA motorcycle that heralded his introduction to tuning and he attempted to raise the compression ratio by bolting a steel plate to the top of the piston. It worked well enough until it melted and the molten metal ran out of the exhaust pipe!

When MG moved to Abingdon, Syd was still working with Morris Garages at the Clarendon in Oxford and was getting restless. He was offered a job with Morris Motors at Cowley, but his foreman, 'Copper' Crease, mentioned him to Kimber, as did Reg Jackson *qv*.

As a result, 23-year-old Enever began work at Abingdon on January 6, 1930, first in the experimental department under Cec Cousins *qv* and soon he became its foreman. Syd was on his way and, above all, was proudest of his essaying EX 135, the Gardner-MG *qv*, which, in June 1939, recorded a speed of 204.2mph in Germany.

He had been appointed MG's chief planning engineer in 1938, but in 1946 he reverted to running the Abingdon experimental department until he was made chief engineer in 1954. Enever was thus responsible for Abingdon's two high-volume models: the MGA and, of course, the MGB.

Enjoying MG is published monthly by the MG Owners' Club, the first issue having appeared in newsletter form in December 1973. Packed with historical and practical information, this highly professional magazine flourished under the editorship of Zoe Heritage and continues to do so under the current holder of the office Richard Ladds.

EX stands for Experiment. Each MG project that originated from Abingdon was allotted an EX number. The register began within a few weeks of the move and the first entry, dated November 6, 1929, relates to the Midget's front wings and lamp supports. During the years from 1935 until 1954 the numbers were allocated by Morris Motors' Cowley drawing office. It had been solely concerned with MG projects until 1953, but EX 178 was applied to '1500cc Austin engine tuning' that reflected MG becoming part of the British Motor Corporation.

When BMC was created in 1952, EX numbers continued to be used, but when a project achieved corporate approval, and thus funding, it also acquired an ADO *qv* number. The MGB, for example, was EX 214 and ADO 23. What was once thought to be the final EX number, 252, is dated March 31, 1977 and refers to rear subframe mountings for the Austin Allegro and Princess. There is a further entry, for EX 253, dated July 3, 1977, but the assignment was not allocated . . . at least for 20 years. Then in 1997 Rover revived MG's famous EX Register and EX 253 was allocated to the turbocharged EX*F* *qv* that ran at Bonneville. EX 254 is the MG*F* Super Sports racer and EX 255 *qv* is destined to be the fastest MG ever . . .

EX 120 was the first of the MG record-breakers, although it did not originate at Abingdon, but with racing driver George Eyston *qv*. A qualified engineer, he was responsible for the Powerplus supercharger and, to publicize the device, he decided to attack the world 750cc hour record. His original intention was to use a little known French car named the Grazide, with a Powerplus in place of its Cozette blower. But testing at Montlhery by Ernest Eldridge, who took the land speed record in 1924, proved disappointing.

Eyston then contemplated using a modified Riley engine mounted in the Thomas Special of 1926 vintage, but he had a chance meeting at Brooklands with an old Cambridge friend, James (Jimmy) Palmes, of MG distributors Jarvis of Wimbledon *qv*. Palmes told him of his plans to attack a number of records then held by the Austin Seven. But his preferred car was the new M-type MG that he was selling.

A meeting was arranged at Abingdon with Cecil Kimber which resulted in the creation of EX 120. Completed in 1930, it was based on the prototype underslung chassis, inspired by that of the French Rally, that was due to appear in 1931 and was destined to serve the MG Midgets until 1955. The attempt on Class H (750cc) records was being planned by George Eyston *qv* and Eldridge.

The factory's Reg Jackson *qv*, working to Eldridge's instructions, built the car. The capacity of the much modified M-type engine was reduced to 743cc by shortening the stroke from 83 to 54mm. It also had a special valve gear incorporating roller bearings, a sprinkling of Bugatti parts and turned down JAP motorcycle valves. At the bottom of the engine was an Eldridge-designed counterbalanced crankshaft. A single-seater doorless green-painted body with cowled radiator and external exhaust pipe was employed.

The car was taken to Montlhery because Brooklands was closed for winter repairs and, on December 30, 1930, Eyston covered 100 kilometres at 87.3mph. But then one of the modified valves broke, although some records were wrested from Austin.

With the supercharged Seven threatening, Kimber was determined that an MG should be the first 750cc car to exceed 100mph. This was clearly the moment for Eyston's Powerplus supercharger and the engine was duly rebuilt and converted.

In February, Eyston broke five records, but could not better 97.07mph. Then, on February 16, he broke a further four and EX 120 attained 103.13mph. The barrier had been broken. In celebration of this achievement MG initiated a production version of EX 120, the C-type Montlhery Midget *qv* .

Later that September, Eyston once again took EX 120 to Montlhery, but having achieved 101.1mph, the car caught fire, the driver ended up in hospital, and that was the end of EX 120.

EX 127, The Magic Midget of 1931, was the second 750cc record-breaker that MG built for George Eyston *qv* and was even more successful than its forebear. It differed from the first car in having an asymmetric rear axle. The engine and drive was therefore offset, which allowed the tall Eyston to sit low in the chassis. The single-seater body was designed at Abingdon by Reg Jackson and a quarter-scale model tested in Vickers wind-tunnel at Brooklands.

With Eyston still in hospital after the fire in EX 120, Ernest Eldridge took EX 127 to Montlhery and achieved 5 kilometres at 110.28mph. Eyston was once more at the wheel, on December 22, 1931, and it covered 114.77mph, breaking four records between 5 kilometres and 10 miles. Now 120mph looked to be within reach.

After a incident-prone attempt at Pendine Sands in February 1932, Eyston made a further attempt at the end of the year and, on December 13 at Montlhery, he attained 120.56mph over the flying-mile and kilometre.

In 1933 the Magic Midget's bodywork was revised to the extent that Eyston was no longer able to fit in, and the diminutive Bert Denly took his place. Austin had already achieved 119mph with its supercharged Seven, but in October 1933 the Magic Midget reached 128.62mph. The year ended with EX 127 having broken all the Class H records.

In 1934 MG sold the car to the German driver Bobbie Kohlrausch, who, after racing it at Nurburgring in the Eifelrennen, found it too small for him. So it was returned to the factory, and early in 1935 the Magic Midget was reborn with a P-type chassis, its third and final body and a Zoller-blown Q-type engine. In this revised form it was allotted the new number of EX 154.

Kohlrausch took the car back to Germany and participated in many races and hillclimbs there. But he could not resist attempting some 750cc records and achieved an incredible 130.89mph over the flying-kilometre and a 130.48mph flying-mile at Gyon, in Hungary.

At the end of the 1935 season, the engine was removed from the car and Kohlrausch's mechanic Artur Baldt returned to Abingdon with a bronze cylinder head made in Germany. There followed ministrations by Baldt, Jackson *qv* and Syd Enever *qv*, and with the engine running on nitro fuel and coupled to a substantial Zoller supercharger blowing at 39psi, the diminutive single-cam 'four' attained an extraordinary 146bhp at 7,500rpm. John Thornley later recalled: "The noise set all Abingdon by the ears." This was the equivalent of nearly 200bhp per litre, and far in excess of any racing engine of its day.

On the Frankfurt/Darmstadt autobhan during the 1936 Record Week, on October 10, Kohlrausch put in an astounding flying-mile at 140.6mph, a record that stood for 10 years, when it was broken by another seemingly indestructible MG record-breaker and driver, Goldie Gardner in EX 135 *qv*.

Kohlbrausch's 1936 achievement greatly impressed Mercedes-Benz, which in 1937 bought the car. In the course of developing the

V12 unit for the M154 Grand Prix racer of 1938, the Magic Midget's engine was tested with the 6⅛in Zoller in place running at 25.5psi, which was higher than that envisaged for the new racer; it developed 123bhp at 7,500rpm on the Unterturkheim dynamometer. This was the equivalent of 460bhp from a 3-litre engine.

Presumably The Magic Midget perished during the wartime bombing on Stuttgart. Does anyone know for certain?

EX 135, the Gardner-MG completed in 1938, was constructed after the company had officially withdrawn from racing and record-breaking. It was therefore officially sanctioned by Lord Nuffield, who was impressed by Goldie Gardner's *qv* MG record-breaking activities in 1937. But this was a cost-conscious exercise and EX 135 was constructed from parts of two K3 Magnettes.

The chassis was based on that of George Eyston's The Magic Magnette of 1934, a made to measure single-seater K3 (K.33023) with offset transmission in the manner of EX 127 *qv*.

It could be fitted with two alternative bodies. The best known, for record-breaking, was finished in MG's distinctive chocolate and cream stripes and duly accorded the name of Humbug. The alternative, titled Coal Scuttle, was intended for road racing events.

In October 1934, at Montlhery, Eyston, driving the car in Humbug guise, took the 200km standing-start record at 128.8mph and a further 11 international Class G records, but he disposed of the car early in 1935 to D N Letts. It was bought in 1937 by Goldie Gardner and went on to form the basis of the rebodied EX 135 record-breaker of 1938.

The engine came from Ron Horton's 1933 Magnette (K.3007), a car which had been fitted with an offset single-seater body by Jensen *qv* in 1934 and which Gardner had bought in 1935 and raced without great success at Brooklands. The engine was Zoller-supercharged from 1936 and Gardner took it to Germany and established a clutch of international Class G records at Frankfurt in October 1937.

The 'new' MG record-breaker was completed in June 1938 and the ex-Horton engine was now Centric-supercharged. An all-enveloping body was designed by Reid Railton, who was responsible for John Cobb's Railton land speed record car. The MG's green-painted body was built by aircraft engineers E G Brown and Company, of Tottenham, London N17.

EX 135 was taken to Germany again and, on November 9, 1938, on the same Frankfurt-am-Main autobahn he had used the previous year, Gardner achieved 186.6mph over the flying-kilometre.

The following year, on May 31, a mere three months before war broke out, he was back, this time on the Dessau autobahn. Intent on

Class G records, Gardner pushed EX 135 to 203.5mph, and on June 2, with the 'six' bored out on the spot to 1,105cc to qualify for Class F status, he attained 204.3mph.

During the war the car was stored in West St Helen Street, in Abingdon *qv*, but some of its components, namely the 1,100cc engine plus two superchargers, disappeared as a result of the fire of 1944. Happily, the car was safe, as was a special N-type Magnette-based 741cc six-cylinder engine prepared for Gardner to challenge Kolhrausch's 1936 triumphs with The Magic Midget.

The MG works was unable to prepare EX 135 for new record attempts in the immediate postwar years, so the all-important engine testing was undertaken by The Stirling Engineering Company, of Dagenham, Essex, which manufactured Chris Shorrock's superchargers for him. He had previously been responsible for the Centric blower that Gardner had used from 1938.

After the war, on October 31, 1946, at Jabbeke in Belgium, Gardner broke the Class H 10-year record held by the Magic Midget and achieved 159.2mph over the flying-mile. In 1948, a 2-litre four-cylinder experimental XK Jaguar engine was fitted and Gardner took a batch of Class E records at 176.69mph, although EX 135 no longer bore the MG badge on its nose

By this time, S V Smith *qv* had replaced Harold Ryder *qv* as the Nuffield director responsible for MG, and he sanctioned a reversion to Abingdon power. On September 15, 1949, with the new 'six' running on three cylinders, resulting in a capacity of 497cc to qualify for Class I, and again in Belgium, Gardner achieved 154.8mph.

Goldie Gardner (in helmet) with EX 135 in Belgium, October 1946.

Spurred by this success, and with only two cylinders operative for Class J (up to 350cc) , the doughty Lieutenant-Colonel was back in Belgium on July 24, 1950, when he attained 121.09mph. This meant that Gardner and MG were supreme in five of the 10 international classes, from 350 to 1,500cc.

In 1951 came another change of engine. This was a Shorrock-supercharged XPAG 1,250cc pushrod 'four', as fitted in the contemporary TD Midget, from which Enever had coaxed 213bhp. The car was transported to the Bonneville salt flats, Utah, USA, and in August a further clutch of records taken.

The year 1952 saw the fitment of an alternative to the XPAG unit in the form of a Marshall-blown experimental 1,973cc Wolseley 6/80 six-cylinder overhead-camshaft VC20 engine. Gardner returned to Bonneville and, on August 16, took some Class E (1,501-2,000cc) records, although the 62-year-old had the misfortune to spin the car and hit his head on a marker post.

Four days later, on August 20, with the blown XPAG reinstated, Gardner attained 202.02mph over the flying-mile. This was destined to be the final competitive appearance by both man and car.

EX 135 had always been its driver's personal property, so MG subsequently bought it from him and this most enduring of Abingdon's record-breakers is now part of British Motor Industry Heritage Trust's collection at Gaydon.

As a result of Gardner's successes with **EX 135** at Dessau in Germany in 1938, he was presented with a beautifully bound book of press cuttings by Prince Richard von Hessen, leader of the NSKK Motor Group. As an MG publication, *The Greatest Achievement of the Year*, noted, this "showed that the record had far more recognition in Germany than in this country".

In retrospect, the starting point of the MGA, this open two-seater, coded **EX 172**, was prepared for George Phillips *qv* to run in the 1951 Le Mans race. Following a second place with his TC in the 1.5-litre class in the 1950 event, Phillips had an automatic entry for 1951. He had already driven the works TDs in the 1950 season and, as he later recalled for *Old Motor* in 1981, he had a productive meeting with John Thornley *qv* at the factory.

"He asked how MG could help for 1951. 'Not a lot', I said, as there was only the TD. He offered to do a more 'posh' job than I had been able to do: what he was offering was a lightened, rebodied TD. I wasn't wild about the idea, I must confess, but I agreed eventually, since there really was very little else they could do."

Based on the production Mark II TD chassis, Phillips was asked by

The MGA's future lines apparent in the TD-based EX 172 of 1951.

Syd Enever *qv* "if I had any ideas about a body? Now I can't draw a straight line from one point to another to save my life, and I suggested something like a lightened, streamlined version of the Sunbeam-Talbot coupe."

But Enever came up with a roadster body clearly inspired by the contemporary Jaguar XK 120. It was built by a service department panelbeater named Wally Kinsey, with formers made by Alec Hounslow *qv* and Henry Stone *qv*. The dashboard, with its octagonally shaped instruments, foreshadowed that of the TF of 1953.

Building the car proved to be a rather subversive operation and, on Tuesdays, it had to be hidden because this was when the Cowley-based director S V Smith *qv* made his weekly visit. Eventually he had to be told and, faced with a *fait accompli*, gave his approval.

As far as engine tuning was concerned, Phillips says that he asked for KE965 valves, which was the type he used in his Le Mans TC . . .

Once completed the car was tested at Boreham Airfield, long before Ford took it over. It was registered for the road and bore the memorable and appropriate registration number UMG 400, courtesy of University Motors *qv*, a sensible move because it did not look the least like an MG!

Phillips went to collect the car, MG contributed a £150 cheque to cover his expenses, he agreed to buy the car for £700 after the race, and then drove to Le Mans in the road-equipped racer.

Unfortunately for George, the entrants suffered from poor quality petrol, *The Autocar* reporting "trouble was . . . visible in some camps, and in most cases was blamed on the fuel supplied." Phillips was dismayed to find that "the MG's engine pinked almost on tickover" and responded by adding another gasket to the cylinder head, so lowering the compression ratio.

As in the previous year he began the driving but, after about four hours, "there was a bang." Investigation in the pits confirmed that the head of one of the valves had come off "and punched a bloody great hole in the top of the piston".

Phillips decided to press on regardless, and his co-driver, Alan Rippon, continued for another couple of hours before retiring on the 60th lap. Petrol had been getting through the hole in the piston and reached the sump, so diluting the oil. As a result the big ends ran.

George was "bloody disgusted. With all the facilities that MG had they couldn't build an engine that would last 24 hours, yet I could do it myself on the proverbial kitchen table. What came out eventually was that they hadn't used KE965 valves - they were just the standard type."

For the company, John Thornley later maintained that the car had performed well in practice. "Clearly he had a class win in his pocket, but two factors prevented this. Firstly, from long experience of the use and abuse of the XPAG engine, Phillips was definitely of the opinion that it was absolutely unbreakable."

George himself confirmed this when he was astonished to discover that there was such a thing as a rev limit, telling his co-driver in 1950, "when you see vapour coming out of the carbs, change gear - that's what I do."

But, says Thornley, with the car lapping in the 80s and achieving 116mph through the kilometre trap down the Mulsanne, "George gave more thought to the Index of Performance . . . than the Class, and saw himself in the big money if only he were to keep his foot down."

After the race the car was no longer driveable and Dennis Herrick, of Armstong shock absorbers, agreed to tow the MG back to Abingdon behind his own car. Unfortunately, on the return journey he had occasion to brake suddenly and the TD hit his vehicle, so it was returned to the factory with a crumpled nose, much to MG's displeasure.

Apart from the engine problems, for Phillips, "the one thing I wasn't too keen on was the fact that you sat on it, not in it". This was the result of Enever having to use a TD frame. To George it was never EX 172 or UMG 400, but 'The Streamliner'.

But Phillips did not buy the car because he believed that MG had let him down. It was retained by the factory and subsequently broken up; it "suffered the fate it deserved", said the driver. But Enever shared the driver's opinion of the seating position, and its EX 175 successor benefited from a brand new chassis-frame. The MGA was on its way.

This was **EX 175.** In 1952, Syd Enever was responsible for what Thornley later described as "UMG 400 put right". Syd overcame the limitations of the TD chassis by designing a completely new and very strong box-section frame in which the driver and passenger sat within the side members, so lowering the floor that was a mere six inches from the road surface. Prudently, he had a spare frame made.

The body was a development of that used on EX 172, although the radiator grille was enlarged, but the variable-pitch full-width windscreen did not reach production. It was also a larger car than 172, which was restricted to TD dimensions. As on the 1951 Le Mans car, a limitation was the height of the XPAG engine, and the bonnet contained a bulge to accommodate it. Bearing the 1952 Berkshire registration number HMO 6, in the works EX 175 was thereafter known as homosex!

Whilst the car was under development, the Nuffield Organisation *qv* merged with Austin to form BMC. But rather than give MG the opportunity to replace the ageing T Series cars, chairman Len Lord *qv* committed the corporation to the Healey 100 at the 1952 Motor Show. This meant Abingdon had to continue with the TD, and its TF replacement was then put in hand.

With George Eyston *qv* wishing to make another record attempt, Syd Enever's first response was to attempt to use EX 175 fitted with a cockpit fairing, wheels spats and an undertray. But wind-tunnel testing revealed its unsuitability, so instead he used the spare frame to create EX 179 *qv*.

For its part, EX 175 was loaned to a brake manufacturer, which crashed what, in retrospect, was the prototype MGA that belatedly entered production in 1955.

The **EX 179** record-breaker of 1954 was, in Syd Enever's words, "like 135 but different". Although George Eyston *qv* had fallen out with Cecil Kimber over the sale of The Magic Midget to Germany in 1934, he returned to the MG fold in 1953. Eyston, who as a director of Castrol spent much of his time in America, was the first to recognize the publicity that accrued from record-breaking activities. By this time, MG was still relying on the ageing T Series Midget for transatlantic sales.

He therefore approached BMC's Leonard Lord *qv* with a view to him driving an MG record car at Utah. But because the Gardner-MG was owned by its ailing constructor, and he had relied on Alexander Duckham for his lubricant, the only answer was to build another car. Crucially he was able to offer some financial backing from Castrol.

Lord agreed, but after some fruitless attempts with EX 175 *qv* in 1953, work began on the first MG record-breaker since 1938. Although Enever had overall responsibility for the project, the majority of the work was undertaken by MG's Terry Mitchell *qv*. It was a front-engined car on the lines of EX 135, but using the spare EX 175 *qv* chassis, powered by the as yet unannounced enlarged 1,466cc version of the XPAG engine used in the TF. Tuned to give 84bhp at 6,000rpm, unlike its predecessors, it ran unsupercharged. Left-hand drive, appropriately, was employed.

The body was similar to that of EX 135, and the car was taken to the Bonneville salt flats. In August 1954, George Eyston and Ken Miles *qv* broke eight endurance Class F records up to 153.69mph and succeeded in averaging over 120mph during a 12-hour run. The now XPEG-designated unit was duly fitted to the America-bound TF 1500 "with record-breaking 1.5-litre engine".

But this was not the end for EX 179. It returned to Utah in 1956, powered by a prototype unsupercharged 1,489cc twin-cam engine, thus qualifying for Class F records. This required that the car be converted to right-hand drive. With Ken Miles and Johnnie Lockett driving, a total of 16 records up to 170.15mph were attained in August.

Later in 1957 came a further visit to the salt flats when EX 179 was fitted with a 948cc BMC A Series engine, tuned to give 57bhp which was to appear in the Austin-Healey Sprite of the following year. This made it eligible for Class G (751 to 1,100cc). On August 13, driven by Tommy Wisdom and David Ash *qv*, it broke three international and no less than 25 American standing-start records. Three days later, in supercharged form, Phil Hill achieved 143.47mph over the flying-mile. This was destined to be the car's final appearance. It was sharing the salt flats with MG's newest record-breaker, EX 181 *qv*. Happily, EX 179 survives in the collection of the British Motor Industry Heritage Trust.

Like EX 179 *qv*, **EX 181**, of 1957, the last of the MG record-breakers, was also initiated by George Eyston *qv* and was created to break the Class F record of 204.2mph established by Goldie Gardner *qv* in 1939. It differed from its predecessors by having its power unit mounted behind the driver in the manner of John Cobb's Railton. This permitted the use of an aerodynamically efficient teardrop-shaped body following extensive experiments by Enever in the Armstrong Whitworth wind-tunnel. It thus became known as

Eddie Maher (left) with Syd Enever (second right) and Alec Hounslow (right) with EX 181. In the background is EX 179.

Roaring Raindrop.

EX 181, although once again overseen by Enever, was, like 179, largely the work of Terry Mitchell *qv*. It had a tubular chassis with the driver sitting ahead of the engine, a Shorrock-supercharged version of the as yet unannounced 1,489cc MGA Twin Cam unit, developing 290bhp at 7,300rpm. Quarter-elliptic springs and a de Dion rear axle, a favourite of Mitchell's, completed an unconventional package. With Stirling Moss at the wheel on August 23, 1957, the car achieved 245.64mph on rain-softened salt, which was over 40mph better than Gardner's figure.

To tackle Class E records in 1959, EX 181's engine was enlarged to 1,506cc and it then developed over 300bhp. On October 2, Phil Hill broke six records and his highest two-way average speed was 254.91mph. This was the last occasion the car was used in anger.

EX 181 then went into retirement at Abingdon, but 18 years later, at the request of Jaguar Rover Triumph's North American dealers, who wanted it to star in a promotional tour, work began in the summer of 1977 on its restoration by MG engineering apprentices Malcolm Green and Martin Dix. The task was completed in October 1978, but a demonstration run at RAF Abingdon proved a sad occasion when driver Jimmy Cox misjudged his braking point and 181 turned turtle.

Fortunately, the car was rebuilt in two months and displayed at the Detroit Motor Show in January 1979. It is now part of the collection of the British Motor Industry Heritage Trust, where it keeps company with its distinguished forebears EX 135 and EX 179.

When **EX 181** achieved its 254.91mph in 1957, the BMC publicity man who wired the news by Telex to Britain did so at around 7pm on Friday October 2. But it was Saturday morning in the UK, so the resulting MG press releases gave the following day of October 3 as the actual date!

It was intended to launch the MGA at Le Mans *qv* in 1955, but body supply problems delayed its arrival until October. Nevertheless, four roadsters, with 18-gauge aluminium bodies - a team of three plus a spare - were built and entered under the **EX 182** name. Although outwardly resembling what was to be the production A, the cars' 1,489cc engines featured special Weslake cylinder heads with twin 1.75in carburettors instead of 1.5in units, and a 9.4:1 rather than an 8.3:1 compression ratio and no cylinder head gaskets. They developed 82bhp at 5,500rpm, as opposed to the customary 68bhp.

One of those tantalizing 'MG might have beens', **EX 186** was a sports-racing version of the MGA Twin Cam that was subversively built in 1959 for that year's Le Mans, but withdrawn at the last minute.

A batch of 10 were planned, but only one was constructed, unofficially, on a modified MGA chassis with a de Dion rear axle. When Ted Lund *qv* saw it at the factory, he described it as "looking like a miniature 300SL Mercedes-Benz". Henry Stone *qv* considered 186 "the most attractive modern MG ever built".

Driven in and around Abingdon by Stone, it was subsequently taken to the Motor Industry Research Association's Lindley circuit, which it lapped at 140mph. On its existence being rumbled by MG's BMC parent, 186 was hastily shipped to San Francisco and owned there by Kjell Qvale, of British Motor Imports *qv*. It is thought to still survive in America.

EX 220 was an experimental Mini-based front-wheel-drive sports car built at Abingdon in 1959 and effectively a scaled-down version of the as yet unannounced MGB roadster. Seen as a possible replacement for the Midget *qv*, it was accorded the ADO 34 designation, with the Austin-Healey version taking ADO 36. A coupe derivative was also built. In 1964, BMC at Longbridge also produced its own Pininfarina-styled ADO 34, which is now part of the British Motor Industry Heritage Trust's collection at Gaydon. But the concept was scrapped because it did not take sufficient advantage of the space-saving attributes of the Mini's transverse engine.

EX 234 of 1964 was intended as a replacement for the MGB and MG Midget. An open four-seater styled by Pininfarina, it was powered by a 1,275cc A Series engine although there was room for the B Series unit. The substructure was the work of Roy Brocklehurst *qv* and Hydrolastic suspension was used. It survives in the Syd Beer collection.

It is hoped that **EX 255** of 1998 will be "the fastest MG ever". To be driven by world land speed record driver Andy Green at the Bonneville salt flats, Utah, USA, in August, it is anticipated that this left-hand-drive MG*F*-based car would exceed the 254.91mph attained by Phil Hill in 1959 at the same venue in EX 181 *qv*. It is powered by a mid-located, longitudinally positioned, twin-supercharged 4.8-litre Rover V8 engine, boosted to a formidable 900bhp.

Andy Green and EX 255 with EX 181 in the background, at the Rover Group's Gaydon engineering centre in May 1998.

EX-E, which is unrelated to Abingdon's experimental EX Register, made a surprise appearance at the 1985 Frankfurt motor show and represented a two-fold objective by BL Cars. It was to display the

EX-E, styled by Gerry McGovern, made a surprise appearance at the 1985 Frankfurt motor show. It represents the starting point of the MG*F*'s lines.

vitality and competence of the company's new studio set up by design director Roy Axe, and represented BL's long-term commitment to the MG marque. Stylistically the work of Gerry McGovern *qv*, later responsible for the lines of the MG*F*, the silver-finished mid-engined coupe was built to accommodate the V6 engine of the 6R4 rally car *qv*, although one was never fitted. On its unveiling the assembled gathering of motoring journalists broke into spontaneous applause.

Project **EX*F*** was the name allotted by Rover to a special MG*F* with extended tail, conceived in 1997 to commemorate the 40th anniversary of Stirling Moss' 1957 runs in EX 181 *qv*. Sanctioned by Rover in June and designated EX 253, it was powered by a 1.4-litre version of the K Series rather than the usual 1.8 unit because its lower compression was more suitable for turbocharging. It developed 329bhp at 7,000rpm. Driven on August 20 by Terry Kilburn, a technician from a Californian Land Rover dealer, it achieved a speed of 217.4mph.

The distribution of MG **exports** in the 1950s and 1960s was the responsibility of BMC *qv*, which was a notoriously unreliable supplier. Ron Lucas, who had previously been Canada-based and was president of British Motor Holdings (USA) Inc, remembered: "With BMC you never got what you wanted where you wanted it. If you ordered 150 green MG Midgets in Vancouver, you were liable to get 120 dark blue ones on the other side of the country in Halifax, Nova Scotia . . ."

George Edward Thomas Eyston (1897-1979) was the racing driver who introduced MG to record-breaking. A member of one of Britain's oldest Roman Catholic families and a direct descendant of Sir Thomas More, Eyston was educated at Stoneyhurst and Trinity College, Cambridge, where he read engineering. He had served in the First World War with the 3rd battalion, the Dorset Regiment and subsequently became a Staff Captain in the Royal Artillery. He retained this 'handle' throughout the inter-war years, and was awarded the Military Cross.

Eyston began his motor racing career after the war and his first notable success came in 1923 when driving an Aston Martin; he won his class at the *Grand Prix des Voiturettes* at Boulogne. Thereafter he raced all manner of cars, most notably Bugatti, Alfa Romeo and Riley.

In addition to his competition career, Eyston was an accomplished engineer in his own right and as technical director of his company, Powerplus Ltd, which had offices at 239 High Holborn, London WC1, and manufactured and marketed his Powerplus vane-type supercharger. It was his need to publicize the device that led to the creation of EX 120 *qv*, and he went on to drive The Magic Midget EX 127 *qv* and The Magic Magnette EX 135 *qv*.

He was the co-driver with Count Lurani in the K3 that won the 1,100cc class in the 1933 Mille Miglia. But at the end of 1934 he fell out with Kimber because of his sale of the Magic Midget to the German Bobbie Kohlrausch, which he regarded as an unpatriotic act.

Eyston's business interests included a directorship of Transport Equipment and other companies associated with the Thornycroft group. He was also a director of Castrol, which led to the construction, in 1954, of EX 179 *qv*, the fourth and last of the MG record-breakers he drove. He also initiated the creation of EX 181 *qv*. Eyston maintained a lively interest in motoring past and present until the time of his death in 1979 at the age of 82.

F

The first of the Wolseley-related six-cylinder MGs, the 1,271cc **F-type Magna** was available in F1 four-seater open and closed forms. The F2 was an open two-seater version and the F3 had four-seater saloon and tourer body styles. The series was produced between October 1931 and December 1932 and a total of 1,250 were built. Chassis numbers ran from F.0251 to F.1500.

The **fastest MG badged car** at the time of writing was the EX 181 record-breaker which, on October 2 1959, attained a speed of 254.91mph at the Utah salt flats, USA, with Phil Hill at the wheel. The fastest production MG is the MG*F* VVC, which is capable of 130mph.

Nicholas (Nick) Fell (born 1959) was the MG*F*'s project director from December 1992 until the car entered production in 1995. After graduating from Imperial College with a degree in mechanical engineering, Fell joined BL in September 1980 as a chassis engineer, a discipline he pursued for the next five years. In the process of taking an MSc degree, as a project he undertook the engine mounting of a three-cylinder petrol engine, and the isolation principles he developed are still used in Rover's current models. With BL at this time closely involved in partnership with Honda from 1985 until 1988, he was resident engineer at the company's factory at Sayama, Japan, which gave him the opportunity of learning Japanese.

On return to Britain he ran the launch team for the 200, 400 and the Honda Concerto and from there went on to run the Rover 200 cabriolet and 200 coupe programmes. There was a brief involvement with the new 400, prior to taking over his PR3 brief. Subsequently accorded the MG*F* name, it was also his mother's initials!

MG's **first racing 'victory'** was not quite the success it purported to be. Staged on the new San Martin concrete circuit outside Buenos Aires, in Argentina, on October 10, 1927, one Alberto Sanchez Cires won a 90-kilometre event at an average speed of 62mph. His car was a distinctly unsporting Flatnose Morris-based MG 14/28 four-seater tourer. It was, however, a race for small cars and, although Cires agreed to drop out before the last lap, he did not do so and went on to win, much to the disgust of his fellow competitors.

MG's **first significant American order** came in the mid-1930s from the Phillip Morris cigarette company. This was for five cars, a P-type and four NAs, that were used to convey a Morris-inspired pageboy named Johnny, but the origin of the cars probably went unrecognized. This was because there were five Johnnies and the MGs were finished in corporate colours of brown and tan with all identifying marks removed so as not to detract from the corporate message.

The association continued after the war when the company again used MGs to convey Johnny. Happily, these appear to have been virtually standard TDs.

Oliver Arkell was probably the **first MG owner** although the name was not applied to cars until 1924. On August 11, 1923, noting the date in his diary, he entered The Morris Garages' Queen Street, Oxford showrooms. Forty-five years later, in 1968, he recalled: "I saw this yellow car in the window. An unusual yellow - the colour of good butter - and it had black mudguards, Mr Kimber was there and he said it was £300." Arkell bought this Raworth *qv*-bodied two-seater Morris Cowley and he did well because, soon afterwards, Kimber raised the price by £50, making it £350, which was twice that of the equivalent two-seater Morris Cowley. He had ordered six bodies from the Oxford coachbuilder and, not surprisingly, they took some time to sell.

The **first four-seater MG** was ordered by G S 'Jack' Gardiner, who had joined The Morris Garages *qv* as a salesman at its Queen Street showrooms in 1923. He later became personnel manager at Abingdon and was secretary of the MG Car Club between 1946 and 1947. What was probably the second open four-seater MG, by Carbodies, went to trials driver Billy Cooper.

The **first MG to be designed specifically for racing** was variously known as the Mark III *qv*, 18/100 or Tigress. Its competition career was, however, shortlived when a single example, driven by Leslie Callingham and H D Parker, ran in the Double Twelve race at Brooklands in May 1930, but its Morris Engines Branch *qv*-prepared engine ingested a piece of carburettor. However, the event was a triumph for MG when the M-type Midget *qv* won the team prize.

MG **first had its own stand,** albeit as The Morris Garages, at the 1927 London Motor Show staged at Olympia, and continued to do so until the 1978 event, held at Birmingham's National Exhibition Centre, where for the first time it shared a display with the Jaguar, Rover, Triumph division of BL. There was an MGB on the BL stand at the 1980 show, even though the car ceased production during the event. MG was absent from the 1981 Motorfair, but returned to the Motor Show in 1982, when the MG Metro and Metro Turbo were displayed. From thereon the marque has regularly appeared at the event, although as part of its parent company's presence.

The **first MG RV8** to be delivered to a private customer in the spring of 1993 was registered K2 MGW, which was the personalized number-plate of its owner, Martin G West, chief executive of the London Scottish Bank.

When MG's Abingdon factory closed in 1980, BL Cars was asked if it would present the MG **flag,** that flew from a mast at the front of the factory, to the town's museum. The company refused, so the *Abingdon Herald* later bought it at the disposal auction for £150 and presented it to the town. It now resides in the collection of the Abingdon Museum.

G

There are no **G** or H-type MGs, despite the existence of the F-type Magna. G was probably avoided for phonetic reasons, would you want to own an 'Emgee gee', but why no H? I and O were skipped to avoid confusing it with the figures 1 and 0, and E was a stillborn project that received the EX 128 designation. U was also disregarded, although what became the MGA was briefly accorded the Series UA designation.

Lieut-Col Alfred Thomas Goldie Gardner (1890-1958) broke his first record in an MG in 1936 and his last, aged 62, in 1952. Always known as Goldie, which was his Scots mother's maiden name, he was educated at Uppingham, started his business career in a broker's office in Ceylon in 1910 and subsequently became a forester in Burma.

Joining up in 1914 as Second Lieutenant in the Royal Artillery, he was awarded the Military Cross in 1916, suffered serious wounds to his right leg and hip in the following year and left the army in 1922 with the honorary rank of Major.

In 1926 he became a director of the garage business Milne and Russell in South Croydon and had begun motor racing in 1924 with a Gordon England Austin Seven. Later, in mid-1930, he got to know Cecil Kimber, "a real enthusiast," and in 1931 he raced a C-type Midget and was the first person to lap the Outer Circuit at Brooklands in a 750cc car.

Driving a C-type, he crashed badly in the 1932 TT, to the detriment of his old war wounds, and therefore decided to confine his competitive activities to Brooklands racing and, from 1936, record-breaking.

In 1934 he was placed third in the 500 Miles race and won the 1,100cc class in a K3 (K.3015) in which he shared the driving with ex-Bentley Boy Dr Dudley Benjafield. But it was the purchase in 1935 of the ex-Horton single-seater K3 that led to the creation of the EX 135 *qv* Gardner-MG of 1938.

Goldie Gardner

Soon after the outbreak of the Second World War, in October 1939 he rejoined his old regiment, the Royal Artillery, as a Second Lieutenant, but was almost immediately promoted to the rank of Major. In 1942 he became a Lieutenant-Colonel and two years later was attached to General Montgomery's staff, serving overseas until demobilization in 1945.

Until his retirement in 1952 he continued to keep the MG name to the forefront in international record-breaking. A measure of the respect he enjoyed, particularly in continental Europe, was recorded at the time of his death, in September 1958, at the age of 68 by *The Motor*.

His activities at Jabbeke made him "a revered figure in Belgian circles . . . where his tall, spare frame, clipped moustache and quiet manner as he stood beside his record car awaiting the results and leaning on a stick to ease a leg wounded in the First World War, endeared him to the enthusiasts as the typical English gentleman."

Goldie Gardner's autobiography, *Magic MPH*, was published in 1951 by Motor Racing Publications.

MG's **Golden Jubilee** was inadvertently celebrated by British Leyland in 1975 when it literally interpreted 'Old Number One' of 1925 *qv* as the first MG. A more accurate 50th anniversary date would have been 1974 or even 1973. It accordingly produced the limited edition Jubilee MGB GT *qv*.

Cecil Kimber always signed his letters in **green ink.** This was, says John Dugdale *qv*, who knew him, a 'green-for-England' gesture.

One of the Rover Group's most experienced engineers, **Brian Griffin** (born 1940), was MG*F* project manager from the programme's 1991 inception until 1992. The son of BMC's technical director Charles Griffin, he joined the Corporation's Longbridge works in 1960 and served a five-year engineering apprenticeship. There he worked in development and chassis design before ending up in the experimental department. He was briefly involved in George Harriman's corporate sports car project, derisively dubbed Fireball XL-5, and project engineer on the Austin 2200 of 1972-75. He subsequently set up a corporate prototype manufacture and build facility and, between 1978 and 1986, managed all the engineering workshops within what had become Austin Rover.

Next he moved to vehicle evaluation at Rover's Gaydon proving ground, and in 1987 became the Group's Canley-based chief engineer, small cars. There he had overall responsibility for the K Series-engined Metro (R8) prior to taking over, on January 31, 1991, the PR3 commission as project manager. He remained in this position until December 1992, when Nick Fell *qv* took over as project director. Brian Griffin continued to work as chief engineer until what had become the *MGF* entered production in 1995.

H

The **Hambro Trading Company** of America Inc became the official importer and distributor of MGs in 1948 and continued to be so involved until 1967. The Louisiana-based business, set up by the Hambro merchant bank, originally imported British china and motorcycles. But an approach by Jocelyn Hambro to Nuffield Exports resulted in the Trading Company importing six TCs that were displayed at Dallas, Texas, and immediately found buyers. As a result, Hambro became the Nuffield Organisation's *qv* sole concessionaires in America.

Soon to be New York-based, the renamed Hambro Automotive Corporation subsequently joined with BMC to create The British Motor Corp/Hambro Inc, with premises in Ridgefield, New Jersey. This continued until 1967, when British Motor Holdings (USA) Inc was established, which mirrored BMC's merger with Jaguar in the previous year. It was, in turn, replaced in 1969 by British Leyland Motors Inc *qv*.

George William Harriman (1908-1973), was chairman of the British Motor Corporation *qv* in the crucial years between 1961 and 1968 and thus chaired the MG Car Company *qv* over the same period.

Harriman began his career with Morris Engines in Coventry in 1923 and rose to the position of assistant works superintendent by 1938. He owed his entire career to Leonard Lord *qv*, who also worked there, and when the latter became managing director of Morris Motors, Harriman spent three years at Cowley during the much needed reorganization. On Lord's departure in 1936, he returned to Coventry.

But when his sponsor moved to Austin, he soon recruited Harriman, who became machine shop superintendent at Longbridge in 1940, production manager in 1944, and a directorship and general works manager title followed in 1945.

With the creation of BMC in 1952 Harriman was promoted to become Lord's deputy chairman and managing director. He succeeded him in 1961 and was knighted in 1965. But this genial, though ineffectual engineer executive found himself out of his depth in the increasingly competitive 1960s. Although not an MG *aficionado*, he opted for an MGC GT as a company car at about the time of BMC's takeover by the Leyland Motor Corporation early in 1968. He was its chairman until November 1968, when he was replaced by Sir Donald Stokes, and Harriman became British Leyland's president, a post he held until his death at the age of 65.

Frank Lachlan Maclean 'Mit' Harris (1897-1945) was the third secretary of the MG Car Club and proprietor of the MG-related magazine, *The Sports Car qv*.

Harris served in the balloon service, spotting submarines, during the First World War and after hostilities he began his journalist career by contributing motoring articles to *The Star*, a now defunct London evening paper. In 1924 he replaced Miles Thomas *qv* as editor of *The Light Car and Cyclecar* and contributed a column under 'The Blower' pen name. Its proprietors, The Temple Press, were sufficiently impressed with Harris to also make him the joint editor of *The Motor Cycle*. Hardworking, outspoken and an enthusiastic caravanner, he resigned in 1932 and the following year founded *The Caravanner*, which still survives as *Caravan Magazine*.

'Mit' Harris

An active participant in competitions, his ambition was to win motor car club events on two, three and four wheels and he was thought to have been the only man to have done so.

A contributor to *MG MaGazine*, in February 1934 he succeeded Alan Hess *qv* as secretary of the MG Car Club and, in 1935, founded *The Sports Car*, of which he was editor until publication was suspended with the outbreak of the Second World War in 1939.

Having served with the RAF in the 1914-18 conflict, Mit Harris returned to that service's balloon command, but later retired, with the rank of Wing Commander, on medical grounds. He died in September 1945 at the age of 48. However, his wife Mary continued to be actively involved in the Club.

Stylist of the MGB roadster, **Donald (Don) Hayter** (born 1927) went on to become MG's last chief engineer. His motor industry career began with Pressed Steel at Cowley in 1942. As this was wartime, the firm was producing complete fuselages and tail sections of Spitfire, Wellington and Lancaster aircraft, and it was Don's job to translate the plans supplied by manufacturers into production drawings.

With the ending of the war Pressed Steel renovated its car jigs and Hayter's brief was to interpret existing schemes and streamline them so that the panels would come off the tools properly.

But Don recognized that he was involved in a late stage of a car's development, and he wanted to be in at the start of the process. Therefore, in 1954 he left Cowley and joined Aston Martin as a draughtsman.

He only remained at Aston Martin until 1956 because the drawing office was to be transferred to Newport Pagnell, and Hayter was not keen to move there. So he wrote to MG on spec and was offered the post of chief body draughtsman.

On arriving at Abingdon in February 1956, his first job was to design the detachable panels in the engine compartment of the MGA Twin Cam *qv* to allow access to be made to the exhaust pipes.

Don's next assignment was the design of the MGA Twin Cam coupe that Ted Lund *qv* ran at Le Mans in 1960. A conversion from an existing roadster, with his Aston Martin background he based it on the DB3S coupe.

By this time the MGA successor was under development and Don and chief body engineer Jim O'Neill *qv* undertook a number of coupe and roadster styling exercises for a possible replacement. His definitive drawing of the MGB tourer is dated June 19, 1958. He remembers, "I took a new EX number, 214, and 214/1 was drawn quarter-size, straight on the paper . . . and that was the MGB all in one go".

Once the car was in production, Hayter was also responsible for the GTS sports-racing coupes (EX 241) of 1967 and the one-off MGB GT SSV1, for System Safety Vehicle One (EX 250), that incorporated existing and projected safety legislation.

Don succeeded Roy Brocklehurst as MG's chief engineer when the latter moved to Longbridge in 1973. As such he was the last holder of the post and remained at the Abingdon factory six weeks after car production had ceased, then left on December 12, 1980.

He was subsequently involved with the Anglicization of the Honda Ballade in Japan that emerged, in 1981, as the Cowley-built Triumph Acclaim, and he retired at the end of that year.

Donald Healey's eldest son **Geoffrey Carroll Healey** (1922-1994) was responsible for the engineering of the Austin-Healey Sprite on which the MG Midget of 1961 was based. Educated at Warwick

School, he studied engineering at the Cranbourne School of Mines and later at the Coventry Technical College.

Following war service with the Royal and Electrical Mechanical Engineers, he left with the rank of Captain and in 1947 rejoined Armstrong Siddeley, for which he had briefly worked earlier. But he soon gravitated to his father's Warwick-based Donald Healey Motor Company, and in 1949 was appointed chief engineer.

After the Austin-Healey Sprite appeared in 1958, BMC decided to produce an MG version which coincided with the arrival of the Mark II Sprite in 1961, that was seen into production by Syd Enever *qv* at Abingdon.

Harry Herring, MG's superlative modelmaker, joined the company in 1930 and remained at Abingdon until he was 73, well after retiring age, such was the quality of his work. The models were created from the styling drawing and were also used for wind-tunnel testing.

For a time, three generations of the Herring family all worked at Abingdon. Harry's son John, known as Jack, went to MG in 1929, almost as soon as it opened there. A coachbuilder who learnt the craft at Caversham Motors in Reading, he began as a door inspector and paved the way for his father, who joined in the following year. Later, in 1950, Harry's grandson John joined MG, having served an apprenticeship with Charlesworth *qv*, and worked for a time in the trim shop and became a welder. His brother Ernie worked as a sprayer in the paint shop, a third son Peter briefly worked at MG, whilst the fourth of the brothers, Tim, worked on the production line and, like Ernie, remained at the factory until it closed in 1980.

Alan Charles Hess (1900-1987) was the second secretary of the MG Car Club in 1932-34 and editor of The *MG MaGazine qv* throughout its publication between 1933 and 1935. Hess was a motoring journalist who also raced and drove at Brooklands on occasions, most successfully in the 1933 Light Car Club International Relay Race, when he ran a trio of stripped MG Magnas driven by himself, Charlie Martin and G W J H Wright, which won at an average speed of 88.62mph. Re-engined, the Magnas ran in that year's Alpine Rally and won the team prize.

Alan Hess was also a writer of fiction and verse, an amateur painter and a broadcaster on motor racing.

During the Second World War he served in the Royal Army Ordnance Corps and later became a sub-editor for BBC home news in 1945 and 1946. Later that year he returned to the motor industry when he suggested to Leonard Lord *qv* that Austin required a public relations officer, and was duly given the job. In that capacity he took part in the company's record-breaking at Indianapolis with the hideous Austin Atlantic in 1949, and in 1951 he undertook a successful attempt to drive around the world in the Austin A40 Sports in 21 days. He wrote books about both initiatives; in all he had nine titles to his credit.

Hess left Austin in 1953 to become director of a portable bailing business, and from there to the Information in Industry organization. But in 1958 he returned as public relations manager for The Simms Motor and Electronics Corporation.

Visitors to the MG factory in prewar days were greeted by **'Topper' Hollis**, resplendent in his brown uniform with cream piping and silver buttons. He came from a long line of porters; his father was so employed at Exeter College, Oxford, and his brother was there for five. Hollis worked there for some 19 years before becoming a club steward, a job he undertook for a further 23. He had joined MG at Edmund Road in October 1928. As the factory's *Broadsheet*, published in 1935, reported: "no-one knows how he acquired his nickname, but he is known through the factory for his cheery smile".

Alec Leslie Hounslow (1911-1976) was MG's immensely experienced development engineer. He joined Morris Garages in April 1925 and initially worked on car assembly. With the move to Abingdon, Hounslow soon showed his worth on racing car preparation and the high point of his career came when he rode as mechanic in the K3 that Tazio Nuvolari drove to victory in the 1933 TT.

When racing came to an end Hounslow became foreman of Syd Enever's development shop where prototypes of new road cars were built and tested. His last job, prior to retirement in 1974, was to design the special inlet manifold on the Rover V8 engine that obviated the need for an unsightly power bulge. He died in October 1976.

I

J S Inskip Inc distributed MG in New York from 1948 until the 1960s. The business had deep-rooted associations with British cars, having been established by Jack Inskip, who had previously been New York sales manager for Rolls-Royce of America and subsequently its last president. With the creation of J S Inskip in 1936, he continued to service Rolls-Royces and coachbuilding activities were perpetuated in the Brewster Building in Long Island City. Coincidentally, Motor Sport Inc, which marketed MGs, was also housed in the same building. Showrooms were also established at 10 East 57th Street, which were occupied by Jaguar after the war.

During the hostilities Inskip moved from both sites and took over premises at 327 East 64th Street. It was from there that, in 1948, Inskip briefly became MG's sole importer in America, having taken over from Motor Sport Inc and Zumbach, but they soon ceded control to Hambro *qv* although remaining a distributor.

It was also responsible for producing some 12 *four*-seater TDs, created by cutting the chassis in half and letting in a new 10-inch long section. The original rear end of the body was retained, but new doors, appropriately secured by Rolls-Royce hinges, were made. Chrome fairings were used along the bonnet sides.

Inskip continued to sell and distribute MGs until 1967, when they ceased retail activities to concentrate on distribution.

Alec Constantine Issigonis (1906-1988), creator of the Morris Minor and Mini, was responsible for the MG independent front suspension that first appeared on the Y-type saloon of 1947. Issigonis joined Morris Motors in 1936 from the Rootes Group, where he had begun to interest himself in suspension design. Soon assisted by MG draughtsman Jack Daniels *qv*, the American-inspired coil springs-and-wishbone system he devised was intended for the projected Morris Ten. But the management opted for a cheaper beam axle that was the work of H N Charles *qv*. Instead Issigonis' design was to be used on the Morris Twelve and the new MG saloon due to appear at the 1939 Motor Show, but cancelled because of the outbreak of the Second World War.

So it did not surface until the Y-type saloon arrived in 1947, and was consequently employed on the TD, TF and MGA and was carried over, in refined form, for the MGB. It thus endured until 1980.

In the meantime, in 1957 Issigonis had been made BMC's chief engineer, and in 1961 he became the Corporation's technical director. Indirectly he was thus responsible for the best-selling MG saloon, based on his popular Morris 1100 of 1962. Less happily, he had overall responsibility for the 3-litre six-cylinder engine used in the MGC.

With the creation of British Leyland in 1968, Sir Alec, as he became in 1969, was made corporate director of research. He retired in 1971.

J

The 847cc **J1** Midget four-seater tended to be overshadowed by its J2 two-seater stablemate. Outwardly similar to its D-type predecessor, apart from cutaway doors, it was built between July 1932 and July 1933. A total of 381 were produced, 262 four-seaters, 117 Salonettes with the two remaining chassis converted to J2s. Chassis numbers ran from J.0252 to J.0631.

The definitive MG prewar sports car, the 847cc **J2** Midget, with its double humped scuttle, cutaway doors and bolster petrol tank, was produced between July 1932 and January 1934. The cycle-type wings were replaced by all-enveloping ones in 1933. A total of 2,083 cars were built, of which 2,061 were two-seaters, some 600 having the later swept wings; 22 chassis were bodied by specialist coachbuilders. Chassis numbers ran from J.0251, J.2001 to J.3750 and J.4101 to J.4432.

Whilst it is well-known that the **J2,** registered RX 9980, road-tested by *The Autocar's* Sammy Davis *qv* in its issue of August 5, 1932, was only capable of achieving its top speed of 80.35mph following 'attention' at Abingdon, neither was it a true J2, but based on the first J1 chassis (J.0251). Many customers complained to MG that their cars could only reach 60mph and were being outperformed by the smaller-capacity but lighter M-types!

The sports-racing 746cc **J3** Midget was effectively a Powerplus supercharged version of the J2 produced between November 1932 and September 1933. A total of 22 were built. Chassis numbers ran from J. 3751 to J. 3772.

The Powerplus supercharged racing **J4** Midget was produced between March and July 1933. A total of nine were built. Chassis numbers ran from J.4001 to J.4009.

The Morris Engines Branch *qv* six-cylinder 2,468cc single-overhead-camshaft **JA**-type engine powered the MG 18/80 family of cars between 1928 and 1933. The work was undertaken by George Pendrell and Frank Woollard (both *qv*). But contrary to past references, MG historian Robin Barraclough believes that it could not have been commissioned by Cecil Kimber *qv* because work on the design began in 1925, long before he would have been in a position to wield such influence. However, The Morris Garages manager immediately recognized its potential and used it in the 18/80, just as he later did with the Morris Minor's Wolseley-designed ohc 'four'.

Pendrell was responsible for the overall design of the unit, whilst Woollard contributed the production engineering element. The JA 'six' is reputed to be the first British engine that did not require its big ends to be scraped prior to fitting.

George Pendrell subsequently set down his reasons for opting for an overhead-camshaft layout in *Automobile Engineering* (New Era Publishing). His requirements for this touring unit were smoothness, silence and excellent low-speed torque. He believed that the extra complication was worthwhile because the head and valves were better cooled than in a typical pushrod design, being placed in two rows either side of the camshaft, rather than in a single row.

The 69 x 110mm engine first appeared in the Morris Light Six announced at the 1927 London Motor Show, some 10 months before the MG 18/80 was announced. The Morris only remained in production until the summer of 1929, when it was replaced by the Isis, which initially was powered by its JB derivative and survived until 1935, some two years after the last 18/80 had been sold. It also formed the basis of the Morris Motors Commodore marine engine.

The JA developed some 60bhp in the MG, and Engines Branch was also responsible for the more powerful Mark III *qv* 18/100 racing version because Abingdon was not yet experienced in such matters. It did not prove successful and the single example withdrew on its 1930 racing debut.

Reginald (Reg) Cecil Jackson (1906-1976), as head of the racing department, was one of the key figures at Abingdon. 'Jacko' as he was known early on, although later he disliked the name, joined MG during its two-year spell at Edmund Road on November 6, 1928. Previously he had run his own small garage at Maidenhead and had worked for GWK in the town.

His first important job was to prepare M-types for their first triumphal appearance in the 1930 Double Twelve race in 1930. He was then assigned the preparation and design, working under Ernest Eldridge's direction, of EX 120 *qv*, the first of the MG record-breakers. Following Eyston's triumph in the car in February 1931, Reg Jackson received a £30 bonus, which was sufficient for him to put down as a deposit on a bungalow in the village of Kennington. He called it Montlhéry, no doubt to the bemusement of the local populace.

Next it was EX 127, the Magic Midget, and from thereon Jackson prepared every MG record-breaker until EX 181's final appearance at Bonneville in 1959. He similarly had overall responsibility for all the Abingdon-prepared racing cars until the closure of the department in 1935.

On occasions he rode as a mechanic to drivers such as Tim Birkin, Eddie Hall and 'Hammy' Hamilton. He would have dearly loved to have partnered Nuvolari in the K3 that triumphed in the 1933 Tourist Trophy as he had supervised the car's preparation. But Cecil Kimber refused to permit it and Alec Hounslow *qv* got the job instead because, said MG's managing director: "Hounslow was expendable, Jackson was not."

From 1935 until the outbreak of the Second World War he was seconded to the service department, where he had a roving commission to solve particularly difficult problems.

During hostilities he worked on munitions production at the factory, and after the war was appointed to the post of chief inspector, a role he undertook until his retirement in 1971.

John Thornley is on record that it was as a result of the efforts of **Dick Jacobs** and George Phillips *qv* on race tracks that he was later able to convince BMC of the need to establish a Competitions Department *qv* at Abingdon. Richard William Jacobs (1916-1987), educated at Branscroft School, Woodford Wells, Essex, became an apprentice with commercial vehicle manufacturers AEC. In 1937 he joined the family business, the Mill Garage in South Woodford, Essex, which had been established in 1919 by his father in what had been the *Old Mill* pub, better known as a local house of ill repute. In 1937 Dick became the proud owner of a British Racing Green J2 Midget.

Following war service as an engineer in the Royal Navy, Jacobs took over the Mill Garage and from 1946 it specialized in MG and Wolseley. He began racing a TC MG (LEV 735), which he stripped and supercharged, but over the winter of 1947-48 he acquired a TA (CS 7695) which was powered by a supercharged ex-army Morris Ten engine and had an aluminium body by John Haesondonck. He first drove it at Goodwood, where he came second to a young man in a Cooper named Stirling Moss. In 1949-50 Jacobs ran in no less than 26 races, finishing in the first five in 18 instances and won on seven occasions. In his first road race, the 1949 Manx Cup, he was placed seventh.

In that year Jacobs, along with Ted Lund *qv* and George Phillips, he was asked by John Thornley to drive a works TC in the first production sports car race at Silverstone. He was placed 19th and soon emerged as the fastest of the trio. For the 2-litre race in 1950, Jacobs, in one of the new TDs, was second, 16th overall, and later took first in class in that year's TT.

This was the last year of Abingdon's direct involvement with racing for the time being, but in 1951, Jacobs, in an ex-works TD, won the 1.5-litre class in the by now familiar sports car race at Silverstone.

In 1952 this event was for production touring cars, and Jacobs won his class three years running from 1952 to 1954 in, of all things, a YB saloon.

When MG returned to racing in 1955, Jacobs shared the driving of one of the EX 182 sports-racers, but had the misfortune to crash at White House. It was the outcome, he later recounted in his book, *An MG Experience*, of the terrible collision near the grandstand and the smoke "must have distracted me and I lost concentration".

He was reported to have died and, in later years, delighted in producing the press cutting to that effect. He switched to the management of two Z Series Magnettes and a pair of Twin Cams for Alan Foster and Roy Bloxam to race. His Abingdon-built lightweight coupes known as 'Jacobs Midgets' scored many class wins in international events, invariably driven by Foster and Andrew Hedges.

Dick Jacobs retired at the age of 58 when, in 1974, his garage business was acquired by compulsory purchase for the building of the M11 motorway. He died in November 1987, a few days before he would have celebrated his 71st birthday.

One of MG's principal distributors, **Jarvis and Sons** of Wimbledon, was run by James (Jimmy) Palmes, a friend of George Eyston *qv*, who remained a Jarvis director until the 1960s. Established in March 1921 at Victoria Cresent, Wimbledon, London SW19, its coachbuilding activities were located at the Grove Works, Merton. John Thornley *qv*, who lived locally, bought his M-type from Jarvis in 1930, so qualifying him as a founder member of the MG Car Club *qv*.

In 1927, Jarvis produced a handsome two-seater duck's back body on the MG 14/40 chassis designed by Arthur Compton, who left the firm in 1929 to establish a body business on his own account.

The firm also marketed an attractive occasional four-seater body on the M-type chassis, first in fabric and then metal, complete with Ducellier headlamps and external three-branch exhaust pipe. It was priced at £255. However, thereafter Jarvis ran down its coachbuilding activities to concentrate on more lucrative car sales. In the late 1960s the business was taken over by Mann Egerton and the long-established Jarvis name disappeared.

Jensen Motors of West Bromwich has MG connections reaching back to prewar days. Established by brothers Alan and Richard Jensen in 1928, their Austin Seven-based Jensen Special featured an open two-seater sports body, which led to a similar offering on a Standard chassis and so to special bodies on Wolseley Hornet chassis.

In an Abingdon context, Jensen was responsible for the offset single-seater bodies commissioned by Thomson and Taylor for Ron Horton's Midget and, in 1934, the K3 Magnette (K.3007). Subsequently bought by Goldie Gardner, its engine was used in his EX 135 record-breaker *qv*.

Jensen was also responsible in 1937 for a handsome open two-seater roadster body fitted on K.3025 for W E C Watkinson, who retained it until 1959, when it was sold to Kjell Qvale, of British Car Distributors *qv*. Fortunately, Watkinson had kept the original racing body, which permitted the factory to remove the later one and replace it with the original. The Jensen tourer was then transferred to the first K2 two-seater (K.2001), a more appropriate location, which shares the same wheelbase as the K3.

Jensen's coachbuilding activities led, in 1936, to limited car production. In 1949, the firm began using Austin engines and, following the creation of BMC, its chairman Leonard Lord *qv* commissioned a prototype from Jensen for a projected corporate sports coupe powered by the 1.2-litre A40 engine.

Lord chose the Healey 100 but, as a consolation, Jensen was awarded the body contract. This business continued until the Big Healey, Abingdon-built since 1958, ceased production in 1967.

The work permitted Jensen to maintain its own car line, but after production ceased in 1970 Kjell Qvale acquired an 85 per cent share in Jensen to produce the Jensen-Healey of 1972, in which MG's former chief engineer, Syd Enever *qv* had a consultative role. Sadly, this lacklustre open two-seater failed to sell in the expected numbers and the 1974 oil price recession was the knockout blow, with the result that Jensen closed its doors in 1976.

The most **jinxed MG** was the prototype SA saloon with an engine capacity of 2,288cc, as opposed to 2,322cc of the production cars. Memorably registered BRM 70, it was run in the 1937 Mille Miglia by *Daily Herald* motoring editor Tommy Wisdom and his wife Elsie. They got as far as Florence, but then skidded on a wet road and hit both a cow and a bridge.

On being returned to Abingdon, the SA was standing on all its Jackall four-wheel hydraulic jacks when they gave way, pinning a luckless apprentice to the floor by his head, happily without serious injury, much to the hilarity of his workmates.

Later, in 1938, Cecil Kimber suffered his first ever car crash at the wheel of BRM 70, driving with his future wife and mother-in-law to Lee-on-Solent. He broke his nose, his wife-to-be had minor cuts and bruises, but her mother's injuries were rather worse; she broke two ribs and received a badly cut head.

Back at Abingdon, the saloon became a works hack, the body was cut down behind the driver and a canvas tilt fitted. It was often harnessed to a trailer, but succumbed to the welder's torch soon afterwards . . .

The **Jubilee Special** MGB GT was finished in green and gold livery and fitted with special V8-style wheels with gold rims, a specification allotted to 750 cars. Announced in June 1975, the price was £2,268, which was £130 more than for the standard car. There was a dashboard plaque on which the dealer could engrave the owner's name and number. In the end, there were 751 of them, after one of the cars was damaged whilst featuring in a television advertisement.

There was also a single **Jubilee MGB GT V8,** completed in June 1975. It was built to the order of David Haddon, managing director of the British School of Motoring. Initially registered 1 BSM, it was later fitted with the appropriate MGB 5 number.

Similarly, a one-off **Jubilee Midget** was built which shared the same livery as the B, but with gold-finished Rostyle wheels. It was decided to raffle it amongst the Abingdon workforce, and this very special Midget went down the line on Monday March 17, 1975, and the detailing was undertaken on the following Saturday. As it happened, the car was won by a shopfloor worker who did not drive, so he decided to sell it to Hartwells of Oxford. There it was bought by a local MG enthusiast, but it changed hands again in 1978 when it was acquired by Michael Cohen, the current owner. It is totally original, apart from the battery, and has just 80 miles on the mileometer.

K

If prime minister Margaret Thatcher had had her way the **K Series** engine that powers the *MGF* would never have been built. As she recounts in her political memoirs, *The Downing Street Years:* "I wanted to cut back BL's investment programme and believed one way of doing this was to buy engines from Honda . . . rather than for Austin Rover to develop its own new engines." Opposing this view was the secretary for trade and industry, Norman Tebbit, who reveals in his autobiography *Upwardly Mobile* that he secured approval in 1985 for BL's "new small engine, which for some time had been in doubt".

Rover's acclaimed **K Series** engine, which is mid-located in the *MGF*, began life in the early 1980s, and its distinctive aluminium sandwich construction was an inheritance from a far more radical experiment, namely a plastic engine that would be thrown away after 150,000 miles of use. Plastic is at its weakest in tension, so BL's Canley-based engineering team evolved a design with long bolts running through the block to take tensile loads. Its less radical although still advanced aluminium replacement was also weak in tension, so the concept was perpetuated.

With emissions regulations and fuel economy ever-present considerations, BL's three-cylinder ECV3 (Energy Concept Vehicle), for which former MG chief engineer Roy Brocklehurst *qv* had responsibility, proved the concept of a pent roof and twin overhead camshafts actuating four valves per cylinder, with plenty of barrel swirl promoted by the rising piston.

With governmental approval, the first experimental K Series unit had run in 1985 whilst the philosophy was first publicly revealed in the O Series *qv*-based M16 twin-overhead-camshaft engine of 1986. All had a 75mm bore, with the 1.4-litre having a 79mm stroke, although a projected single-overhead-camshaft 1.1-litre K8 version had a stroke of just 63mm.

The twin-cam 16-valve engine, developing 94bhp at 6,250rpm, first appeared in the Rover 200 of 1989 and was later extended to the

Metro of the following year, which was also offered in K8 guise.

When what became the *MGF* was beginning to evolve, the 1991 mid-engined PR3 concept was powered by the Metro's 1.4-litre twin-cam K Series unit, although early experimental cars used the 1.6-litre version being developed for the new 400 model. However, the creation of a V6 derivative for the 800 saloon allowed engineers to initiate stepped 'damp' cylinder liners that permitted a larger bore. The result was the 80 x 89mm 1.8-litre 118bhp 'four' used in the MG, the 200 and the Land Rover Freelander.

The most complex model range in MG's history is the six-cylinder **K Series Magnette** family of 1932-34 vintage. There were two chassis lengths, the K1 having a 9ft wheelbase and K2 a 7ft 10$\frac{3}{16}$ in one. In addition, three different engine options were available, the 1,087cc triple-carburettor KA, the twin-carb KB and the 1,271cc KD. The KA-engined K1 was usually produced in saloon form but was also available as a four-seater touring model with KB engine. The latter unit, or a KD version, also powered the open two-seater K2.

About 71 examples of the K1 were built between October 1932 and July 1933. Saloon and tourer chassis numbers ran from K.0251 to K.0321, the latter being produced between February and July 1933. Simultaneously about 15 KB-engined K2s were built from K.2001 to K.2015. Perhaps five KD-engined K2s were made between July and December 1933, running from K.2016 to K.2020. Some 80 KD-engined K2 saloons, from within the range K.0322 to K.0439, were assembled between July 1933 and March 1934.

The **K3**, MG's most famous and successful racing car, was a 1.3-litre six-cylinder supercharged open two-seater produced between April 1933 and October 1934. Thirty three examples were built, including two prototypes (K.3751/2) and EX 135 *qv*. Chassis numbers ran from K.3001 to K.3031.

The **K3** was the first production racing car in the world to be fitted with a preselector gearbox. ENV provided MG with special Type 75 units which were used in the three K3s that ran in the 1933 Mille Miglia. Brian Wilson, younger son of the box's inventor, Walter Wilson, later recalled: "The boxes ran perfectly and we had an eulogistic letter from Lord Howe, which was published in full in the motoring press. From thereon every sporting enthusiast wanted Wilson boxes, and we fitted many."*

Paradoxically, the 'box had first appeared as an option in the decidedly unsporting Armstrong Siddeley in 1929, but MG's application followed its use by Malcolm Campbell in the Sunbeam Tiger which he ran in the 15-lap mountain handicap at Brooklands in 1932.

**Walter Wilson: Portrait of an Inventor*

There was little in **Cecil Kimber's** early life to suggest that he would go on to found one of the world's greatest sporting makes. Born on April 12, 1888, his father ran a Manchester-based printing machinery business, and Cecil attended Stockport Grammar School and then Manchester Technical School, where he studied accountancy at evening classes. He later placed budgetary control as the most important of managerial skills.

After an apprenticeship with his father's business, he married in 1915 and later that year joined the Sheffield-Simplex company as assistant to chief engineer A W Reeves. In 1916 the pair were responsible for presenting an extraordinarily progressive paper to the Institution of Automobile Engineers on Works Organization.

But he soon left, was briefly with AC Cars, and then a stores organizer for Martinsyde Aircraft, at Woking, Surrey. In 1918 he joined axle and gearbox manufacturer E G Wrigley, of Birmingham, as "an organization expert" and it was there that he met Frank Woollard *qv*, who went on to co-design the single-overhead-camshaft engine used in the MG 18/80. Kimber had invested all his savings in Wrigley, and when the firm experienced financial difficulties Cecil and his wife left the Midlands in the depression year of 1921 and moved to Oxford with their baby Betty. A second child, Jean, was born in 1925.

Kim, as his intimates knew him, had been made assistant general manager of The Morris Garages in the city, an appointment which paved the way for the creation of MG. He was due to work for Morris, in varying capacities, for the next 20 years.

Cecil Kimber's sacking by Lord Nuffield *qv* in November 1941, alas, followed a pattern of the latter losing talented and able executives who had helped to make Morris Motors the largest car manufacturer in the country. An unhappy period, first with Charlesworth Bodies *qv* and then Specialloid Pistons, culminated in his death on February 4, 1945, when he was one of two passengers to die when the train in which he was travelling ran backwards outside King's Cross station.

To George Tuck *qv*, who so effectively handled MG's publicity for 10 years, **Cecil Kimber** was "quite an enigma, he wasn't an engineer, he wasn't a salesman, he wasn't an accountant."* But the creator of the MG make, slight of stature, just 5ft 5in tall and with a right leg some two inches shorter than the left, the result of a motorcycle accident in his youth, was one of the British motor industry's most talented executives.

His multi-faceted mind possessed an extraordinary eye for line and form, which was coupled with a natural born gift for organization and order. He was, in John Thornley's words, "a visionary".

Cecil Kimber

Kimber also possessed a flair for choosing talented subordinates, and one of his many successful appointees was Thornley himself, who succeeded him. The same went for his choice of MG's outstanding chief engineer H N Charles *qv*. When asked about how he went about designing a car, Charles' answer was always the same. "First of all it is necessary to find someone whose enthusiasm is so infectious that he can persuade others to follow him." The "someone" was Cecil Kimber; the description also fitted Thornley to a tee, but there was a downside to his imaginative, inspired enthusiasm.

To Miles Thomas *qv*, MG's general manager was a "short wiry man, who walked with a limp . . . and took periodic wiggings for lack of profit making with impudent sang-froid, confident that prestige in sporting events by the MG car one year would be recouped in sales the following". Thornley has similarly recalled "he was not a terribly good administrator and was constantly having to go to Lord Nuffield and ask for money".

Kimber's life was dominated by a series of extraordinary interventions that could be regarded as fateful. His motorcycle accident of 1910 effectively put an end to his life on two wheels and resulted in him taking up motoring. The unexpected departure of the general manager of The Morris Garages in 1922 provided another link in the chain of events that culminated in the creation of MG. And in 1945 came his own tragically premature death in a freak railway accident.

The stylistic abilities he so clearly possessed were probably inherited from his mother, who was an accomplished watercolourist. They are ably displayed in the Morris Garages' first MG Super Sports of 1924, particularly in its ingenious application of proportion and colour that is to be found in all his subsequent MGs. The J2 Midget of 1932 represents the acme of his genius. But how did he go about it?

George Tuck has provided an insight of how Kimber worked during the Abingdon years. "Kim kept this drawing block in his office at a big table near his desk; he was always drawing bold lines, and when Harold Connolly *qv* came down to the works they sat at this block and played around with ideas . . . I wouldn't say that's how all our bodies were designed, but the general character and outline of the thing was basically Kimber, and the development by Harold."*

So, like his great contemporary, the equally enigmatic William Lyons of Jaguar, Cecil Kimber was a chief executive who was responsible for styling his products. Thornley has said that Kimber "was determined to out-Jaguar Jaguar [SS]" and, paradoxically, both MG and that Coventry company remain amongst the handful of British makes to have obtained true global status and survived into the 21st century.

**The Kimber Centenary Book*

Cecil Kimber's annual salary more than doubled in the years between 1934 and 1938. In the former year it stood at £1,250, but by 1937-38 had risen to £2,750. The motor industry was regarded as good payers in the inter-war years. Miles Thomas *qv*, Kimber's opposite number at the much larger Wolseley concern, was in 1937 earning £6,000 per annum.

Cecil Kimber's interests, as he recorded them for *Who's Who in The Motor Trade* for 1934, were "trout fishing, sailing, work", in that order!

The **KN Magnette saloon**, with K1 chassis and N-type 1,271cc six-cylinder engine, was produced between July 1934 and September 1935, powered by the N-type engine. Chassis numbers ran from KN.0251 to KN.0451.

Richard (Dick) C Knudson is America's best-known MG enthusiast. A former associate professor of English at the State University of New York at Oneonta, he was co-founder, in 1964, of The New England MG T Register. The first of his many MG books, *The T Series MG*, was published in 1973, and he was responsible for *MG: The Sports Car America Loved First*, and *MG: The Sports Car*, both of which appeared in 1975.

Dick had a sabbatical year at Abingdon in 1976, when John Thornley was a near neighbour, which stood him in good stead when he came to edit *The Kimber Centenary Book*, published in 1988. He also ghosted Henry Stone's *MG Mania: The Insomnia Crew* that appeared in the same year.

L

Unlike its K Series stablemate, that was produced concurrently, the six-cylinder **L-type Magna** was only available in 1,087cc form. A total of 478 were produced in open four-seater, Salonette and Continental coupe versions. It was manufactured between January 1933 and January 1934 and chassis numbers ran from L.0251 and L.0736.

The L2 was only built in two-seater form, 90 being produced between March 1933 and January 1934. Chassis numbers ran from L.2001 and L.2090.

Leslie (Les) James Lambourne (born 1931), was MG's director and general manager from 1969 until 1972. Educated at Bicester Grammar School and Oxford College of Technology, Les Lambourne joined Morris Motors in 1947 as an engineering apprentice. At its completion he became process planning engineer and in 1955 assistant works manger.

In 1958 Lambourne moved to MG, where he was appointed supplies manager, he became assistant works manager in 1964 and works manager two years later. He deputized for John Thornley during his spell of assistant general manager in the 1967-69 era, when he was promoted, by which time MG was part of British Leyland *qv*.

The **last Midget** (G/AN6 229526) was built on Friday, December 12, 1979, one of a batch of 500 Commemorative cars that were also timed to coincide with the 50th anniversary of MG's presence at Abingdon. They were finished in black with a tan interior, although black upholstery was available as an alternative. A plaque on the dashboard recorded that it was a 50th Anniversary Midget from Abingdon.

The **last Midget** carried a *Save Abingdon* sticker on its windscreen, together with the message "gone but not forgotten" that was also attached to the boot. A makeshift black-painted coffin carried the MG badge and the initials *R.I.P.*

The **last MG** to be produced at Abingdon was a pewter metallic LE *qv* MGB GT, G/HD6 523002, that left the line on Thursday, October 23, 1980, and is owned, like the last Midget, by the British Motor Industry Heritage Trust. The factory closed its doors on the following day.

Although officially the **last MGA Twin Cam** (YD1-2610) was completed on April 14, 1960, arch-enthusiast Michael Ellman-Brown obtained John Thornley's agreement to build one additional car, and some four weeks later, on May 18, work began on YD1-2611, which really was the last of the line. The roadster was completed on June 14 and collected by its delighted owner, who still owns it. Ellman-Brown is the author of *MG Collectibiles*, published in 1997.

The **last American-specification MGB**, a black roadster, was bought by Henry Ford II for the Henry Ford Museum at Dearborn. It was very appropriate because his father Edsel *qv* was probably the first American owner of an MG, having taken delivery of an M-type Midget in 1930.

The **last non-MG** to be produced at Abingdon was the Vanden Plas 1500, a model that was trimmed there from August 1979 until the autumn of 1980. This was necessary because this ungainly Austin Allegro-based saloon had been produced in small numbers at BL's Vanden Plas plant at Kingsbury, North London, since 1974 and that facility closed in 1979. Despite its title, most examples were 1,750cc-powered!

The **last occasion** that an Abingdon-designed and prepared MG appeared in a motor race was on August 21-23, 1968. Two examples of the MGC-related GTS coupe, coded EX 241, were entered in the 84-hour three-day *Marathon de la Route* at Nurburgring. One, MBL 446E, driven by Vernaeve, Hedges and Fell, finished sixth and the other GTS retired with overheating. They were subsequently sold off with a dismantled third car, together with the remaining three alloy/steel bodies.

The **last YB chassis**, YB 1551, formed the basis of a glassfibre-bodied coupe raced by Dick Jacobs *qv*. Completed early in 1954 and memorably registered XNO 1, its registration number was subsequently changed to 982 VWL.

MGs first ran at the **Le Mans** 24-hour race in 1930 and appeared there on 19 occasions until 1965. The 1930 event was not an auspicious start, both M-type Midgets, one driven by Francis Samuelson and MG's Freddie Kindell and the other by R C Murton Neal and Jack Hicks, retired. Samuelson and Kindell tried again in a C-type in 1931, but were again unsuccessful, as were the Hon Mrs Chetwyn and H H Stisted in another C. In 1933, a blown C-type driven by Ludovic Ford and M H Baumer succeeded in finishing and managed sixth place. This encouraged them to try again the following year, this time in a K3, but they had the misfortune to end up in a ditch after avoiding a slower car. However another K3, driven by Charlie Martin and Roy Eccles, finished fourth, which was the best ever performance by an MG in the race. They also won the 2-litre class. Ann Itier and Charles Duruy in a PA were 17th.

In 1935 the French entrants Maillard-Brune and Druck, also in a K3, owned by chocolate millionaire Jacques Menier, came in ninth and won the 2-litre class.

This was the year of the George Eyston's Dancing Daughters *qv* all-lady team of Joan Richmond/Margaret Simpson (PA1711), Doreen Evans/Barbara Skinner (PA1661) and Margaret Allan/Barbara Eaton (PA1677), who finished, respectively, in 24th, 25th and 26th positions.

MG was now out of racing and there was no event in 1936, but in 1937 MG was represented by Dorothy Stanley-Turner and Joan Riddell in a PB, who came 16th. It was the same story in 1938 when Mme Itier returned and, with C P Bonneau, finished 12th.

The 1939 race attracted two MGs, those driven by Miles Collier and Lewis Welch and the Belgians Bonneau and Mathieu, but neither featured in the results.

In the first postwar race in 1949, George Phillips *qv*, drove a special-bodied TC, and in 1950 he was second in the 2-litre class and placed 18th overall. He was less fortunate in 1951 when his works-prepared EX 172 *qv* dropped a valve. It was his last appearance at the Sarthe circuit, and another four years were to pass before an MG again appeared there.

The make's first official return to the race tracks for 20 years occurred in 1955 when a team of three prototype MGAs, under the EX 182 *qv* coding, were entered. Johnnie Lockett and Ken Miles *qv* were best placed, bring their car into 12th place. Ted Lund *qv* and Hans Waeffler were 17th, although Dick Jacobs *qv*, who shared the wheel with Joe Flynn of the third car, had the misfortune to crash.

But the event was overshadowed by the death of over 80 spectators in a horrific crash. MG's BMC management subsequently withdrew from such competition and no cars appeared there until 1959. Then Ted Lund and Colin Escott, with quasi-works support, entered a twin-cam roadster, but did not finish.

They were luckier in 1960 when the car, converted to a coupe, was 12th and won the 2-litre class. In 1961 Lund was joined by Bob Olthoff, but they did not complete the race.

No MGs were run in 1962, but for the following three years MGBs were prepared by the BMC Competitions Department *qv*, although in the first event the car with a distinctive extended nose was entered in the name of driver Alan Hutcheson, who shared the driving with Paddy Hopkirk. They came 12th and won the 2-litre class, despite having to dig the car out of sand at Arnage.

In 1964 Hopkirk teamed up with Andrew Hedges, and they were 19th on this occasion in a similar car, having averaged 99.9mph for the 24 hours.

In 1965 the same team again succeeded in completing the 24-hour classic and came in 11th, which was the best position attained by MG in the postwar years. It was an appropriate note on which to bow out.

When the MG*F* was under development, project director Nick Fell *qv* and his team briefly contemplated entering a example in the production car class at **Le Mans**, just prior to it entering volume production in 1995. But wisely, they decided against it. Fell recalls: "I think it was right to hold back because it would have occurred right at the point when we'd got all the pressures of launching a new car."

Designer of MG's famous octagonal badge *qv*, **Edmund (Ted) John Frank Lee** (1903-1986) attended Oxford High School, where mathematics was his best subject and he also excelled at art. He joined The Morris Garages *qv* in the early 1920s, destined for a career in cost accountancy, and Cecil Kimber *qv*, who recognized the importance of the subject, arranged for the business to pay half his tuition fees. A fellow of the Institution of Company Accountants, Ted Lee introduced an American cost accounting system to The Garages. He designed the MG's badge *qv* in about 1923 and moved with Kimber to Abingdon, but later transferred to Cowley and became group chief accountant for all Morris' companies. Lee recalled getting Cecil Kimber to take out a General Accident life insurance policy, with a 2s 6d (12.5p) premium, just before his death. He retired from what had become BMC in the 1960s.

Left-hand drive was not applied to a production MG until 1948, when the Y-type saloon was so equipped and a limited quantity were built until 1951. The first MG sports car to be offered with the option of left-hand drive was the TD of 1950. This was a reflection of the make's growing appeal in America.

The MG with the greatest **length** was the WA saloon of 1938/39 vintage that measured 16ft 2in overall. The shortest was the M-type Midget, which was a mere 10ft 3in long, the same as the WA's wheelbase!

In 1932 C H S Supplies, of Great Saint Andrew Street, London WC2, was selling, for 45s (£2.25), a chrome-plated **lighter** in the shape of an M-type Midget shod with miniature Dunlop tyres. It was activated by pressure on the steering wheel. This was a contemporary of Magic Midget lighters variously made in silver, pewter and chrome on brass.

The first MG to be available in a **Limited Edition** came in 1967, when only American customers were offered "The First Anniversary MGB GT Special . . . only 1000 . . . will be available." Extras included an official plaque, 16-inch wood-rimmed steering wheel with "matching Australian coach wood shift knob" and racing-type wing mirrors.

In 1979 came a second US-designated **Limited Edition** of the MGB roadster to commemorate 50 years of production at Abingdon. Announced at New York's Expo Auto and finished in jet black with silver stripes and British flag motif logo, they were also distinguished by a front spoiler, wider alloy wheels of the Triumph Stag type, a stainless steel luggage rack and a dashboard plaque. Not quite as limited as its predecessor, no less than 6,682 were produced between March and December 1979.

The third and, as it transpired, the final **Limited Edition (LE)** version of the MGB went on sale, but to UK customers only, in January 1981, which was three months after production of the car had ceased. Built between August and October 1980, in roadster and GT forms, there were actually 1,001 examples, but only 999 of these were offered for sale because the British Motor Industry Heritage Trust retained the last of each version. The roadster was finished in bronze metallic paint with solid-colour LE side livery and an orange and brown interior. The coupe had pewter metallic livery and silver side livery with a silver and grey interior. In all 579 GTs and 420 roadsters were offered for sale.

Founder of Motor Racing Publications, the publishers of this book, **Nevil Lloyd** (1912-1962), raced one of the Bellvue Garages' K3 Magnettes (K.3015) in 1936, having been at school, Brighton College, with that establishment's co-owner, Kenneth Evans. In 1946, shortly after the war, Lloyd established MRP, which in 1948-49 was based at 15 and 17 Stert Street, Abingdon, just down the road from MG.

Longbridge, to the south-west of Birmingham, was for many years the Austin factory and today, as the Rover Group's largest manufacturing plant, it is where the *MGF* is manufactured. When Herbert Austin left the Wolseley *qv* company in 1905, where he had been general manager, he went to inspect a disused printing business at Longbridge named White and Pykes. This was to become the Austin factory and, unlike Morris' works at Cowley, which was essentially an assembly operation, Longbridge was much more of a manufacturing plant.

In addition to the mainstream Austin cars, this is where the MG's rival racing Sevens were prepared, but Longbridge did not begin to figure in MG history until the creation, in 1952, of the British Motor Corporation *qv*.

Both the A and B Series engines *qv* that powered the MG Midget and the Z Series Magnette, MGA and MGB were designed and manufactured at Longbridge.

It was in the East Works that Howard Dancocks and Harry Dewick undertook the initial work on the B Series MGB engine to modify it to meet American emissions requirements.

The first MG-badged cars to be assembled at Longbridge were some of the 1100/1300 family, although the overwhelming majority were built at Cowley. The plant was also the location of the assembly of the MG Metro 6R4 Group B rally car in 1984-85.

At Longbridge, the *MGF* is being produced in the revived CAB 2 plant, built by BMC in 1962 for the manufacture of the Mini and 1100. It had last been used for the Rover 200 and related Honda Ballade, which ceased production in 1989, and had thereafter been mothballed. It was refurbished as a niche product facility, and the MG shares the production line with the previous-generation Rover 200 cabriolet and coupe and the 400 Tourer estate car.

Leonard Percy Lord (1896-1967) was associated with MG throughout his working life, initially in 1933-36, when he was managing director of Morris Motors *qv*. As chairman of BMC between 1952 and 1961, he also chaired the MG Car Company *qv*. Coventry-born Lord was educated locally at Bablake School and served his apprenticeship with Courtaulds in the city. He joined

Hotchkiss in 1922 – the business that the following year became Morris Engines – as a draughtsman in the jig and tool office and later had responsibility for machine tools. Destined to become one of the country's outstanding production engineers, when Morris experienced difficulties with the Morris Minor's overhead-camshaft engine, which was also used in the M-type MG, Lord was moved to Wolseley, where he rapidly introduced a cheaper side-valve engine for the model in 1931.

Morris was impressed, and in 1933 Lord arrived at Cowley, where he replaced Edgar Blake as managing director. He then embarked on a £300,000 refurbishment programme and was responsible, in 1934, for introducing the Morris Eight that was the best-selling British car of the 1930s.

But Lord, tough, decisive and blunt, was a flawed personality. His one-time colleague, Miles Thomas *qv*, described him as having an "inverted inferiority complex", whatever that might be. For although capable of individual acts of kindness, he invariably feared rather than respected ability. Once in a position of power, he could exercise it with irresponsibility and malice that often resulted in an individual's dismissal or resignation.

Riley racing driver Adolf von der Becke, who served an apprenticeship at Cowley, has said that Leonard Lord had little time for Cecil Kimber *qv*, a prejudice that was reinforced when he attempted to introduce a sporting car at Cowley. The resulting Morris Ten-Six Special of 1934 was not a success.

As part of his reorganization of Morris Motors, Lord convinced Lord Nuffield that MG and Wolseley, which he personally owned, should be bought by Morris Motors, an initiative that came into effect on July 1, 1935. A visit to Abingdon confirmed his worst fears. When he saw the innovative, newly minted R-types in the racing shop he characteristically commented: "well that bloody lot can go for a start".

Lord had a point. MG had a patchy financial record in the buoyant mid-1930s, so racing ceased, the design office closed down and was transferred to Cowley. But Morris Motors' managing director clashed with Lord Nuffield over his share of profits; he left in 1936, muttering vengeance, and in 1938 joined the rival Austin company.

His vendettas continued, and such was his hatred for the recently deceased young and able Murray Jamieson, who had essayed the exquisite twin-cam Austin single-seater racers that he also loathed, in 1940 he instructed his deputy George Harriman *qv* to disable their engines by disposing of vital parts in a wartime scrap metal drive. This would ensure that they would never run again.

The unique crankshafts and connecting roads were duly retrieved, but Harriman hid one set in his garage. There they remained until after his death in 1973, when they were discovered by his widow and reunited with one of the cars that ran again in 1974 for the first time in 35 years. Lord was not a man to cross, and he held MG's destiny in his hands.

When Austin combined with the Nuffield Organisation to form, in 1952, the British Motor Corporation, he became chairman, a position he held until his retirement in 1961.

Nicknamed Lord Foulmouth by the BMC workforce, Leonard Lord's antipathy to Morris Motors, "those buggers in the country", was underlined, in 1952, when he created a rival make to MG, the Austin-Healey, which left Abingdon with the prewar-based T Series cars.

He also went out of his way to make life uncomfortable for such able and talented engineers as MG's H N Charles *qv*, who left Austin in 1946, while Morris' A V Oak and Alec Issigonis *qv* moved in 1952 because of the anti-Nuffield prejudices that emanated from Longbridge. But the latter returned to the corporate fold late in 1955 and went on to create the revolutionary Mini Minor.

In 1956, Lord summarily dismissed his tough but able potential successor, Joe Edwards, over pre-lunch drinks. This left BMC with his pliant and ineffectual protégé, George Harriman, who succeeded Lord as BMC's chairman.

Happily for MG, one of the people with whom Lord could communicate was John Thornley, who has since reflected: "I must have been one of the few people who did not get the wrong side of his tongue", and it was this positive dialogue that permitted MG's articulate and diplomatic general manager to get the Abingdon drawing office reopened in 1954. Had this not been the case the plant would have surely succumbed to closure.

Lord, who had been knighted in 1954, retired in 1961 and became Lord Lambury in 1962. He died on September 13, 1967, at the age of 71. His retirement had been spent farming, and he enjoyed oil painting. But the results of the damage that he inflicted on the British motor industry are being felt to this day.

A semi-official MG works driver, along with Dick Jacobs and George Phillips, **Edward (Ted) Walter Kingstone Lund** (1917-1975) put his involvement with the make down to the fact that his father had attended Stockport Grammar School with Cecil Kimber *qv* and they were both members of Manchester Motor Club.

He began motor racing in 1947 at the wheel of a supercharged PB, registered BTU 260, and following a successful first season, he entered it in the *Grand Prix des Frontiers* in Belgium, where he was placed fourth in his class.

Ted Lund (right) reunited with his MGA Twin Cam in 1974. With him is the car's owner Bob McElroy.

In 1949, Lund, along with Dick Jacobs *qv* and George Phillips *qv*, was invited by John Thornley *qv* to drive works TCs in the first production car race at Silverstone, when he was placed 17th. In the following year's event he was third in the under-2-litre race. At the TT, MG won their class in which Ted was placed third and came 20th overall.

His own PB was rebodied, and in 1950 the family garage business of E K Lund, of Coppall, near Chorley, Lancashire, was appointed MG agents. It became a popular meeting place for MG Car Club members in the north-west.

He continued his racing career at the wheel of a much-modified TD, registered GBL 412, and later a stripped TF. With the reopening of MG's racing department in 1954, Ted was invited to drive one of the factory's EX 182 sports-racers *qv* in the following year's Le Mans *qv*, and he competed there again in his MGA Twin Cam in 1959, 1960 and 1961.

Ted was reunited with his old car at the MG Car Club's Silverstone race meeting staged in 1974 to celebrate the marque's 50th birthday.

Barré Lyndon (1896-1972) was the first writer of MG history, although he was clearly happier with fiction than fact! The name was a pseudonym and concealed the identity of a prolific journalist and playwright named Alfred Edgar Frederick Higgs. Although he wrote for boys' magazines under the name of Alfred Edgar, he adopted the hero of Thackeray's 18th century Irish gentlemen of fortune for his first play, *Speed*, which adventurously had a motor racing theme and was staged in 1931.

It was a subject he perpetuated for the first of his triology of MG books, *Combat* (William Heineman, 1934), which mostly chronicled the company's racing programme, launched in 1930, and followed it through the 1931 and 1932 seasons. Next came *Circuit Dust* (1934), and *Grand Prix* (1935), both under the John Miles imprint. But MG's withdrawal from competition in mid-1935 meant that there could be no more. All the books were also produced in de luxe editions that were limited to 100 for the first two titles and 75 for the third.

A friend of Cecil Kimber, Lyndon also wrote promotional material for the MG, co-operated with George Eyston for *Motor Racing and Record Breaking* (Batsford, 1935) and contributed to the *The Sports Car* and *The Autocar* on MG-related subjects.

Meanwhile, he was enjoying considerable success as a playwright. In 1937, two of his plays were running simultaneously in London, *Hell-for-Leather*, also with a strong racing element, at the Phoenix and *They Came by Night*, at the Globe.

Lyndon was fortunate that his 1936 thriller, *The Amazing Dr Clitterhouse*, featuring Ralph Richardson in the title role, was made into a film by Warner Brothers in 1938, starring Edward G Robinson and Humphrey Bogart.

Soon afterwards, Barré Lyndon moved to America to become a film script writer in Hollywood and his first screenplay was of his book *Sundown*, set in First World War Africa and released in 1941. Thirteen more followed, the last being *Dark Intruder (Something with Claws)* of 1965. Having made his home in Beverly Hills, he died in America in October 1972 at the age of 77.

M

The first of the Midget family, the 847cc **M-type** was the most popular sports car in the world of its day. Built between March 1929 and June 1932, a total of 3,235 were built. Of these, 2,329 were the fashionable fabric pointed open two-seater with the 273 balance consisting of metal-panelled cars. There were 493 fabric-bodied Salonettes, 37 metal-panelled coupes, Double Twelve racers and replicas accounted for a further 21 cars, and there were 82 chassis bodied by coachbuilders. Chassis numbers ran from M.0251 to M.3485.

The Morris Minor chassis, and thus the **M-type's frame**, was designed under the direction of Percy Rose, the chief designer of Morris Commercial Cars, at the factory where Cecil Kimber *qv* had once worked. After serving an apprenticeship with Royce and Company, the forerunner of Rolls-Royce, in Manchester, Rose later moved to Crossley in the city and in 1922 joined W R M Motors at Cowley. When Morris Commercial was established at the former Soho, Birmingham, works of E G Wrigley, for whom Kimber had worked in 1918-21, Rose became chief engineer in 1937.

The 1,598cc **MG Maestro 1600** hatchback, BL Cars' LM 10, was available between March 1983 and October 1984. Chassis numbers ran from SAXXCTWY7AM 100217*, Y indicating 1.6 litres. But the R Series *qv* engine ran unhappily with the bespoke twin Weber 40 DCNF carburettors. In April 1984 it was superceded by the 1.6-litre S Series unit, courtesy of the Montego, that was fitted until October when it was replaced by the 2-litre O Series *qv*. A total of 12,398 R Series cars were built and just 2,762 1.6-litre S Series ones.
**Early number.*

The MG and Vanden Plas versions of the 1983 **Maestro** were distinguished by the presence of electronic digital instrument displays, and the package included a synthesized voice which reminded drivers to "please fasten your seatbelt", of "low fuel" or "handbrake on." In all there were 32 words of instructions, the voice being that of actress Nicolette McKenzie, who had appeared on such television programmes of the day as *General Hospital*, *Callan*, *Van Der Valk* and *Triangle*. Her voice emerged from the driver's-side radio speaker, but the gimmick was soon discontinued.

The MG Maestro 1600 was replaced in October 1984 by the 1,994cc MG **Maestro 2.0 EFi,** powered by the O Series engine *qv*. It was produced until October 1991 and chassis numbers were prefixed SAXXCTWU7AM, with U indicating 2 litre, and ran to 686179. A total of 32,725 were built.

The 1,994cc MG **Maestro Turbo** was produced between January and November 1989, but continued to be listed until 1991. Chassis numbers ran from SAXXCTWT7AM, the first and second Ts indicating MG and Turbo respectively, 575026* and 623669. Just 505 cars were built.
**Early number.*

The **Magic Magnette**: see EX 135.

The **Magic Midget**: see EX 127.

When, in 1974, the Abingdon-based Morland brewery opened its first new pub in the town since the war, it was named, by popular demand,

Morland's *Magic Midget* pub upholds MG honours in Abingdon.

The Magic Midget. Located in Preston Road, not far from the factory site, the car is depicted on the exterior of the building in its 1933 record-breaking form.

The Longbridge-designed, Cowley-built 1,489cc MG **Magnette Mk III** saloon had no Abingdon imput and was the MG version of BMC's Farina saloon. With just 16,676 examples manufactured, the Mark III was a resounding flop. Chassis numbers ran from GHS1 101 to 16776. Its 1,622cc Mark IV successor lingered on for a further seven years and was discontinued by British Leyland. Chassis numbers ran from GHS2 16801 to 31120. A mere 14,320 were built.

Edward (Eddie) Joseph Maher (1910-1976), Morris' Engines Branch's *qv* talented experimental engineer, spent most of his early working life at Riley, which he joined in about 1929. He soon graduated to the experimental and racing department and was largely responsible for the cars so effectively campaigned by Percy McClure. He was initially unaffected by Morris' takeover of Riley in 1938, but with the outbreak of war Maher was seconded for development work on the SU carburettors used on the Rolls-Royce Merlin aero engine.

On rejoining the company after hostilities, he developed the 1.5 and 2.5-litre engines for production, but later moved from the Riley factory, at Durbar Avenue, to Engines Branch *qv*, at Courthouse Green, Coventry. There he became development engineer, where he worked in tandem with chief engineer James R Thompson. He saw the MGA Twin Cam engine into production and similarly applied his talents to the BMC A Series engine *qv* first used in the Austin-Healey Sprite and later the MG Midget. Despite being crippled with polio, he subsequently became chief engineer and retired in 1974. He maintained a lively interest in the Coventry marque and was president of the Riley Register at the time of his death in September 1976 at the age of 67.

John Thornley's, ***Maintaining the Breed***, inspired by a corporate advertising slogan, was published by Motor Racing Publications in 1950. Written with good humour, perception and authority, it chronicles MG's racing and record-breaking activities from a factory standpoint and is the only instance of a British motor industry executive having a book published whilst in office.

Now usually known as the 18/80, but variously called the **Mark I** or the MG Six, this 2,468cc car in the Bentley idiom was produced in saloon and four-seater touring forms, and with bespoke coachwork, between August 1928 and July 1931. Chassis numbers ran from 6251 to 6750 and 500 were built.

The first of the Abingdon-designed MGs, the chassis numbers of the **Mark II** were accordingly prefixed with the letter A. Produced in open two and four-seater, saloon and coupes forms, it was introduced in April 1930, concurrently with its Mark I predecessor, and was made until October 1932. A total of 236 were built and chassis numbers ran from A0251 to A0486.

The sports-racing version of the Mark II, the **Mark III** 18/100, had purpose-designed engine components, dry-sump lubrication and uprated brakes. Produced between October 1930 and January 1931, just five were built. The chassis number ran from B0251 to B0255.

Only one MG was ever offered with a **mascot**. This was the Mark III *qv* that was unofficially known as the Tigress or Tiger, although the mascot may depict a panther! Sometimes ascribed to Gordon Crosby *qv*, the leaping big cat has recently been identified by MG authority Roger Stanbury as the *Panthere*, the work of the French sculptor Casimir Brau. As only five cars were produced, the mascot is accordingly extremely rare. Because of its weight it was mounted on the car's substantial headlamp bar rather than the radiator cap.

Cecil Kimber had one, plinth-mounted on his office desk, as did publicity manager George Tuck *qv*. William Morris' example complemented a chrome-plated ashtray and, significantly, is engraved *C. Brau* on its base. This may lend credence to the suggestion that six replicas of an original were made by MG at the time and some rough bronze castings of the mascot remained at the factory for many years afterwards.

The tiger as featured in an MG advertisement of February 1930.

A similar big cat featured on an MG advertisement for the 14/40 model in 1928 and, two years later, a rendering found its way on to the Mark III's sales literature. And in February 1930, Kimber told *The Motor* that the impending Sports Six would, "in all probability be called the MG Tiger". But there was no reference to the name when the car was introduced three months later!

The Crosby association has probably sprung from the fact that he had an example mounted on the radiator cap of his own 18/80 saloon and because of its similarity to the Jaguar mascot he created for the SS Jaguar range in 1938. That is unless you have any further ideas . . .

Mayflower Vehicle Systems, formerly Motor Panels, not only manufactures the MG*F*'s body, its parent company is also a partner with the Rover Group *qv* in the project. The Mayflower Corporation, founded as recently as 1989, has an unexpected parentage. On one hand was John Simpson's Timelaw group, of which Ribbons, the car seat belt company, was the best known ingredient. On the other was the Triangle Trust, in effect a shell company, chaired by merchant banker Rupert Hambro, that once produced Tri-ang toys. Coincidentally it was a subsidiary of Hambro Bank that, in 1948, took over the distribution of MG on the American market.

Simpson and his partners sold the business to the Trust, which was renamed The Mayflower Corporation, and in 1991 it bought the well-

MG*F* bodies under construction at the Coventry works of Mayflower Vehicle Systems

established Coventry bodybuilding business of Motor Panels. Located in the city's Holbrook Lane and formed in 1920, it specialized in low-production body supply and had worked for most of Coventry's car makers over the years. This included SS, as Jaguar was known in prewar days, its next door neighbour between 1928 and 1952, and it even owned the business in 1939-44. It was then bought by Rubery Owen, who sold it in 1989 to the CH Industrials Group, which by this time owned Tickford *qv*, the company that was responsible for the completion of the low-volume MG Maestro Turbo *qv*. CHI's origins were rooted in one-time MG supplier, the Coventry Hood and Sidescreen Company, but it succumbed to a recession-induced collapse in 1991, which was when Mayflower bought the business.

Rover's cost-conscious PR3 project, that flowered as the *MGF* in 1995, required a partner, and in 1993, Mayflower was not only awarded the body contract, but also agreed to invest £24.2 million in the new car. Destined to run for six years, Mayflower Vehicle Systems, as Motor Panels became in 1996, produces the bodies-in-white, to a maximum of 20,000 a year, at its Coventry premises. These are then dispatched by Foden lorry, eight at a time, to Rover's Longbridge *qv* factory. There they are sprayed in a new £42 million waterborne paint plant, prior to delivery at CAB 2.

Frederick Wilson Henry McComb (1927-1989), was editor of *Safety Fast qv* from 1959 until 1964 and author of *MG by McComb*, the standard work on the marque. An Ulsterman educated at Friends' School, Lisburn, Northern Ireland and Queen's University, Belfast, McComb began his journalistic career as motoring correspondent for the Northern Ireland Region of the BBC. He contributed local coverage for the weekly motoring magazine *Autosport qv*, of which he became assistant editor in 1953 and remained until 1955. He left the magazine to become a technical author with Castrol prior to taking over the editorship of the *Autocourse* racing annual in 1957. It was from there that John Thornley *qv* recruited him to become founding editor of *Safety Fast*, a post he filled with distinction. This gave him the opportunity to interview longstanding members of the MG workforce, many of whom had worked for the company since its Oxford days. He was also BMC competitions press officer in the years between 1964 and 1967. In 1963, McComb had taken over as general secretary of the MG Car Club *qv*, which was run from the factory, but on the withdrawal of works support he left in 1969 to become a freelance writer, broadcaster and photographer. In 1972, Dent published McComb's first book, *The History of the MG Sports Car*, that in 1978 was expanded and enhanced with photographs as the acclaimed *MG by McComb* (Osprey) that is still in print at the time of writing (1998). He had updated the work in 1984 and also wrote *MGB* (1982) and *MGA* (1983) for Osprey. His *Aston Martin V8*, and *Mercedes-Benz V8* in the same AutoHistory series display the same readability and meticulous research that characterized his MG titles. Wilson McComb died aged 62 from a longstanding heart condition.

The individual most closely associated in the public mind with the MG*F* is its stylist **Gerry McGovern** (born 1956). After education at Binley Park School, Coventry, he joined Chrysler UK's young designer development programme in 1973 and obtained a degree in transportation design at what is now Coventry University. He subsequently obtained a master's degree in the same subject at the Royal College of Art. Moving briefly to Chrysler's Detroit headquarters in 1978, he returned to Coventry that year and to Peugeot Talbot, which had bought Chrysler's UK operations. As principal designer, he was recruited in 1982 by Roy Axe to join the

Gerry McGovern with MG*F* owner Anthea Turner on the occasion of the model's first birthday party in 1996.

new Austin Rover concept design studio established at the former Triumph car factory at Canley. There he was responsible for the MG EX-E concept car and the CCV (Coupe Concept Vehicle) of 1986, prior to joining the PR3 programme in 1991. He was subsequently responsible for the lines of the acclaimed Land Rover Freelander. Unlike many of his breed, the extrovert McGovern, with his distinctive shoulder-length hair, is prepared to talk in depth about his work, the quality of which speaks for itself.

The arrival of the 1,275cc MG **Metro** hatchback in May 1982 marked the welcome reappearance of the MG name. A 100mph version of the successful Austin Metro, BL's LC8 of 1980, it was available until September 1990. Chassis numbers approximately began at SAXXBGND1AD, with G indicating MG, 551595. A total of 120,197 were built.

When the MG **Metro** was under development, BL's planners experimented with an instrument panel that featured octagonal instruments, but dropped the idea because it visually jarred with the rest of the car.

The potent but fragile 1,275cc MG **Metro Turbo** was introduced in October 1982 at chassis number SAXXBZND1BDV640054 (Z indicates Turbo) and was available until September 1989. A total of 21,968 were built.

The mid-engined four-wheel-drive 2,991cc MG **Metro 6R4** of 1984-85 was created by Williams Grand Prix Engineering for what proved to be the shortlived Group B rally championship that was discontinued in 1987. Some 200 examples were built.

Registration numbers with an **MG** prefix were an ingenious and memorable enhancement to those cars supplied by the company's sole London distributor University Motors *qv*. Issued by Middlesex County Council, the firm was allocated, between March 1930 and March 1949, the numbers MG 1 to 8000. In the meantime, in June 1933, the AMG prefix had been reserved by Middlesex for other outlets, but in May 1949, by which time the two-letter plates had been exhausted, University were allotted the series from UMG to YMG, which ran until May 1953. Later, in March 1959, it took over a batch in the reversed UMG sequence, but booming sales meant that they were soon used up and 999 UMG was issued in January 1960.

The sought-after **MG 1** registration number that was originally fitted to a University Motors *qv*-supplied MG was discovered, in 1958, on a Vauxhall by BMC competitions manager Marcus Chambers *qv*. The car was bought by MG, its number then graced an early MGA Twin Cam (YD1 524), was transferred to a factory MG Midget and eventually found a home, in 1973, on John Thornley's new MGB GT.

The **MG Car Club,** established in 1930, was for many years of its existence factory-based, although today it is an independent organization and still flourishes at Abingdon.

It was M-type Midget owner Roy Marsh, of Highbury, London, who had the idea of a club for MG owners. He put pen to paper and his letter was published in *The Light Car and Cyclecar* of September 5, 1930. Marsh noted that "a number of one-make clubs have been formed lately with satisfactory results. Now Midget enthusiasts, what about an MG car club?".

On October 12, an inaugural gathering took place at the Roebuck Hotel, near Stevenage, on the Great North Road, and *Light Car* journalist Harold Hastings later recalled that "over 340 MGs turned up - all but two of them Midgets - and the car park looked rather like a dispatch bay at the works".

The chairman was Stanley Kemball, sales director of University Motors *qv*, who subsequently offered his office for meetings. The hon secretary was John Thornley *qv*, then a young City of London- based accountancy student, who agreed to approach Cecil Kimber for the club to use the MG name and badge. Thornley later recalled that "he was enthusiastic and even commissioned F Gordon Crosby *qv* to design the club badge".

But demand was such that Thornley soon realized something would have to give, and it was not going to be his commitment to MG! This led to him joining the company as deputy service manager in 1931 and running the club from the factory. But his work there meant that, in September 1932, he stepped down in favour of Alan Hess *qv*, although he continued to be actively involved in events.

The following year Hess also took over the editorship of the club's first journal, *The MG MaGazine qv*. By this time, regional centres had been established and the first overseas one set up in Ceylon, whilst the formative American centre was New York-based.

Alan Hess stepped down as secretary at the beginning of 1933 and his place was taken by F L M 'Mit' Harris *qv*, who later became editor of *The MG MaGazine's* successor *The Sports Car qv* and ran the club from the publication's Holborn offices.

With the outbreak of the Second World War, Harris rejoined the RAF, *The Sports Car* ceased publication and club news appeared, first in *The Motor*, and then *The Light Car*. MG's sales manager Hector

Kimber House, the MG Car Club's headquarters in Cemetery Road, Abingdon.

Cox become caretaker secretary and, in doing so, the Club's running once again reverted to the factory. With Harris' untimely death in 1945, the job was taken over by personnel manager Jack Gardiner, although John Thornley replaced him at the end of 1947. But the disruption of the war had taken its toll and membership had dropped to 385.

Happily, by the beginning of 1951 it was nearing the 2,000 mark when Liverpool-based Russell Lowry, northern editor of *Autosport qv* magazine, took over as secretary, although the factory continued to provide support.

In August 1947, the club had held its first race meeting at the new Silverstone circuit, and the membership graph continued to accelerate; it reached 5,000 in 1955 and exceeded 6,000 by late 1957.

The demise of *The Sports Car* meant that the club was temporarily without its own magazine, and it wasn't until 1959 that Thornley, by then MG's general manger, recruited Wilson McComb *qv* to Abingdon to run *Safety Fast qv* magazine. With Lowry's retirement in 1963, McComb also took over as club secretary

By 1968, membership had risen to 5,810 in eight centres throughout the country and some 3,000 spread through 50 centres the world over. But the event that was destined to have a far-reaching impact on MG, the Abingdon factory and the MG Car Club occurred with the creation of British Leyland. The net result was that, in 1969, the club began an independent existence with Gordon Cobban, secretary of the South Eastern Centre, as secretary. Although no longer factory-based, its presence was maintained in the town of Abingdon.

By this time separate registers were well established, and these have been maintained over the years. There are currently 14 of them, the most recent addition being for *MGF* owners.

Racing has always been an important ingredient of Club activities, and meetings are held at Silverstone, Brands Hatch, Snetterton, Oulton Park and Cadwell Park.

Membership currently stands at some 12,000 and the MG Car Club's address is Kimber House, PO Box 251, Abingdon, Oxon OX14 1FF. The full-time chief executive, appointed in 1997, is Robert Gammage, formerly vice-chairman.

The **MG Car Company** (Proprietors the Morris Garages Ltd) was registered in March 1928 and was followed by the incorporation, on July 21, 1930, of the MG Car Company Ltd to take over the car manufacturing arm of The Morris Garages *qv*. Sir William Morris *qv* was governing director, in other words chairman, and Cecil Kimber *qv* managing director. The business remained Morris' personal property until July 1, 1935, when it was purchased by Morris Motors *qv*. With the creation of the British Motor Corporation *qv* in 1952, the MG Car Company Ltd continued to exist as a wholly owned subsidiary of Morris Motors and just survived the 1968 takeover of BMC by the Leyland Motor Corporation to form the British Leyland Motor Corporation *qv*.

MG Cars is the Rover Group *qv* subsidiary established in 1995 to handle the sales and marketing of the *MGF*. Dealers were required to underline their commitment to the marque by producing a business plan and, in consequence, 120 in the UK were awarded the MG franchise. Initially Longbridge-based, MG Cars now operates from the Group's Bickenhill offices, near Birmingham's National Exhibition Centre, which also houses the Rover and Land Rover sales companies.

MG enthusiast Magazine first appeared in 1983 and has been edited from the outset by Martyn Wise. Initially available bi-monthly, since 1993 it has been published monthly by MG Enthusiast Ltd, of PO Box 22, Dewsbury, West Yorkshire WF12 7UZ. Regular contributors include Andrew Roberts and Malcolm Green.

The MG MaGazine, subsidized by the company, first appeared on a bi-monthly basis in May 1933. It was published by the House of

The first magazine to be solely devoted to the make survived for two years.

Simpson (Publishers), with offices at 418-422 Strand, London WC2. It was produced under the editorship of Alan Hess *qv*, who combined his duties with the secretaryship of the MG Car Club *qv*. It thus carried club news together with many related MG items, and survived until March 1935 (Volume Two, No 12). There was also a dummy issue, prepared for prospective advertisers. It was replaced by *The Sports Car qv*.

There would be a gap of 44 years before the title was revived in 1979 by John Dugdale *qv*, the national product publicity manager of British Leyland Motors Inc *qv*. In this instance there was no definite article, it was *MG Magazine*, but it only survived with corporate backing until 1981 when MGB sales ceased. Since then it has been produced independently and remains so at the time of writing (1998).

With some 46,000 current members, **The MG Owners' Club** is the world's largest one-make car club. It is the brainchild of the dynamic Roche Bentley, who started it in 1973 after experiencing difficulties in obtaining spare parts for his black 1966 MGB roadster.

At the time he was a London-based office equipment salesman with Pitney Bowes, and Bentley has recounted its origins in the club's monthly magazine *Enjoying MG qv*. Twenty-three years later he remembered: "Fed up with the hopeless service in parts and information and advice, I had mused that it would be a great idea to start an MG Club to help what must be hundreds of people like me to run their MGs a bit more easily."

In October 1973, after prudently registering the name MG Owners' Club, Roche placed an advertisement in *Exchange and Mart*. "It basically offered information on a new club for MGs and it invited response."

He's never looked back. The need for such an organization was underlined by the sackloads of mail that were delivered to the club's first office, the back bedroom of a semi-detached house in Luton. Before long, Bentley was promoting "the services of specialist traders who had sprung up to sell and repair MGs all over the country . . . and although I didn't know all the answers I did know someone who did."

Such was the growth in the business that, in 1977, Roche was able to give up his job and work full-time for the MG Owners' Club, in which he was joined by his brother Martin and friend Richard Monk. That same year the club moved to new headquarters in rural Cambridgeshire. However, this wooden studio in the grounds of a 16th century thatched cottage proved too small, and in 1979 the club's administration was transferred to rented offices in the village of Swavesey.

Then, in 1979, BL announced that it was going to close MG's Abingdon factory and production of the MGB was to cease. Bentley's "Save Abingdon" campaign elevated the MGOC to national prominence, which was reflected by a growing membership that soared from 11,000 to 22,000 in a single year.

The MG Owners' Club was in the forefront of keeping the MG flame burning, and in 1982 came yet another move, this time to larger premises that were opened by Viscount Lindley, a keen MGB owner. By this time there were 18 full-time staff on the payroll.

The MG Owners' Club's purpose-built Octagon House, in Station Road, Swavesey, Cambridgeshire.

During the 1980s, membership continued to soar to the extent that, by 1989, the organization had again outgrown its office accommodation. But that year British Rail offered for auction the former railway station goods yard at Swavesey, that had since become a coal storage depot. The club bid for the site and it was successful. On January 1, 1992, work began on clearing the weed-infested land of coal and rubble, and six months later the new purpose-designed headquarters, Octagon House, was up and running. It was subsequently opened by Stirling Moss.

In addition to administrative offices and shop, there is a workshop facility, where members can leave their cars to be attended to by a fully trained staff. There is also a body shop, and the club also supplies its members with spare parts and insurance services. It publishes a Directory of Recommend Suppliers which is based on members' countrywide experiences.

There are some 140 Club areas within Britain, and events for MG enthusiasts are held on practically every weekend of the year. Membership is open to the owner of any MG, and that, of course, includes the MG*F*. The MG Owners' Club's address is Octagon House, Swavesey, Cambridge CB4 5QZ.

The first recorded instance of the use of the **MG name** appeared in a Morris Garages advertisement in *The Oxford Times* of March 2, 1923. Displaying a range of Morris, Humber and Sunbeam cars, it encouraged potential buyers to "order now for Easter". In the border surrounding the copy are six representations of the MG octagonal badge, the work of the Garages' cost accountant Ted Lee *qv*.

MG Six: see Mark III.

The Morris Garages' handsome **MG Super Sports 14/28**, based on the 1,802cc 'Bullnose' Morris Oxford, was built between late 1924 and August 1926. Invariably bodied with four-seater touring coachwork by Carbodies *qv*, it was also available in open two-seater, four-seater saloon, landaulette and Sports Salonette forms. Chassis numbers fell into the Morris range: (1924) 35,001-57,000, (1925) 57,001-103,500, (1926) 103,501-168,000. A total of 336 were produced.

The **MG Super Sports 14/28** name was perpetuated on The Morris Garages' 1,802cc Flatnose-based replacement produced from late 1926 until August 1927. Available in open two and four-seater and two and four-seater Salonette forms, 290 were produced.

It was replaced by the outwardly similar, still Morris-based, but MG-badged 1,802cc **14/40 MG Sports Mark IV**. Minor visual differences from the 14/28 included the introduction of an apron below the radiator to conceal the dumbirons, and the deletion of an extended base to the radiator. Chassis numbers ran between 2251 to 2765, and 486 were built between November 1927 and June 1929.

Most recent of the MG magazines, the bi-monthly ***MG World*** appeared in 1997, edited by Philip Raby and is published by CH Publications, of PO Box 75, Tadworth, Surrey KT20 7XF. Noted MG historian David Knowles is a contributor, as is the author, who as The History Man, also writes a regular column.

Abingdon's first volume seller, the 1,489cc **MGA** roadster, was produced between May 1955 and May 1959. A total of 52,478 were built and chassis numbers ran from G/HD 10101 to 68850. The coupe version, of which there were 6,272, ran from September 1956 until May 1959 with numbers running from HM 20671 to 68850.

The 1,588cc MGA 1600, introduced in May 1959, was produced until March 1961. There 28,730 roadsters, from chassis G/HN 68851 to 100351, and 2,771 coupes, G/HD 68851 to 100351.

The 1,622cc Mark II MGA was produced between March 1961 and May 1962. Roadster chassis numbers ran from G/HN2 100352 to 109070 and coupes from G/HD2 100352 to 109070. There were 8,198 roadsters and 521 coupes.

Although the rear dickey seat was popular in the 1920s, incredibly, a company offered such a conversion on the **MGA** 1500 in 1957. John Gibson and Son, of George Street, Edinburgh, transferred the spare wheel to a rack on the bootlid that was converted to hinge at the bottom rather than the top. Squabs were fitted to the boot floor and the lid and it was possible, said Gibson, to accommodate two people. The cost was £67.10s (£67.50), but were any sold?

The **MGA** had the most successful export record of any British car. Of the 101,081 produced between 1955 and 1962, no less than 91,547 were exported, which represented 91 per cent of total output. The overwhelming majority, 81,153, went to America.

In an attempt to increase the popularity of the **MGA**, the factory briefly contemplated producing it in four-seater form. Although allocated the EX 189 designation, it seems that the idea came to nothing. There is, however, a very professionally converted four-seater MGA in existence, imported from Singapore, although its origins are unknown. There were also thoughts about producing a cheap version of the A, which carries the EX 195 designation in the company's experimental register.

The 1,588cc **MGA Twin Cam** was produced in roadster and coupe forms between April 1958 and May 1960, making a total of 2,111 cars, 1,788 open and 323 closed ones. Chassis numbers ran from YD501 to 2611.

The **MGA Twin Cam coupe** first used by Ted Lund *qv* at Le Mans in 1960 was finished in the same metallic green paint used on the EX 181 *qv* record-breaker. This is because the one-off factory-designed body was built by Midland Sheet Metal, of Chilvers Coton, Nuneaton, Warwickshire, which had also been responsible for EX 181's body and had some paint left over from its construction and decided to use it up!

The Mark I 1,798cc **MGB** roadster, designated EX 214 by Abingdon and ADO 23 by BMC, entered production in May 1962 and was built in this form until October 1967. Chassis numbers ran from G/HN3 101 to 138360, making 115,898 cars.

Its Mark II successor was built between November 1967 and September 1969. Chassis numbers began at G/HN4 138401 and 31,767 were produced.

Not quite the last MGB leaves the Abingdon production line in October 1980.

The G/HN5 'Mark III' Bs were built between September 1969 and September 1974, from chassis 187170, the final chrome-bumpered roadster being G/HN5 359169. A total of 110,643 were built.

The 'rubber'-bumpered roadsters were produced between September 1974 and October 1980 and ran from G/HN5 360301 to 523001, making 128,653 cars.

The **MGB GT** was MG's EX 227 and built between September 1965 and October 1967. Chassis numbers ran from G/HD3 71933 to 137795, making 21,835 Mark I GTs produced.

The Mark II, built between November 1967 and September 1969, ran from GHD4 139471 and amounted to 16,943 cars.

The chassis numbers of the chrome-bumpered G/HD5 series, produced between September 1969 and September 1974, ran from 187841 to 360069 and amounted to 59,459 cars. The 'rubber'-bumpered GTs built between September 1974 and October 1980, from G/HD5 36001 to 523001, totalled 27,045.

The 250,000th **MGB** was a GT built on May 27, 1971, and was used as a prize in an American competition, The Great 250,000th Giveaway. It attracted no less than 70,000 entries and the winner was one Bill Newton, of Mobile, Alabama.

The 3,528cc **MGB GT V8**, MG's EX 249 and British Leyland's ADO 74, was marketed between August 1973 and July 1976. Chassis numbers ran from GD2D1 124 to 2903 and 2,591 were produced, of which seven were experimental cars for a possible entry to the American market.

The **MGB GT V8's V8 radiator badge** was courtesy of the contemporary Rover 3500 Mark II saloon, both of which were powered by essentially the same engine.

The 2,912cc six-cylinder **MGC**, BMC's ADO 52, was built between November 1967 and August 1969. Chassis numbers ran from G/CN1 (roadster) 138 to 9102 and (GT) G/CD1 110 (November 1966) to 9102.

The **MGC UMS**, which stood for University Motors Special, came from a batch of 141 cars, 118 GTs and 23 roadsters, bought by longstanding MG distributor University Motors *qv* in 1969, after production had ceased. They were finished in special paintwork and had a bespoke radiator grille and badging. Their engines were uprated by Downton Engineering's associate London-based company, Richard

Miles (Downton, London), of Elvaston Mews, Queens Gate, London SW7, in Stage II or III states of tune. But not all Downton-converted Cs were UMS cars. There were also cosmetic enhancements, but no two were alike and, ultimately, only 21 cars were so modified. It was not until 1971 that the last of the heavily discounted cars were sold.

The most famous **MGC customer** was HRH Prince Charles, who took delivery of his GT in 1969. Registered SGY 766F, it was stabled in the Royal Mews in Buckingham Palace and he was often seen in it at Windsor polo matches. It is currently exhibited at the museum at Sandringham.

The **MGC's engine**, the controversial six-cylinder Type 29G unit, was so designated to reflect its MG destination and cubic capacity. Designed at Morris' Engine Branch *qv* to Alec Issigonis' *qv* brief, it was an uneasy combination of BMC's C Series *qv* unit and BMC Australia's 2.4-litre B Series *qv*-derived 'six'. It thus inherited the C's 83 x 88mm internal dimensions and 2,912cc, but differed in having its camshaft and valve gear on the right-hand side, permitting the use of separate valve ports rather than siamesed ones. The outcome was an engine that weighed a substantial 650-700lb and developed 145bhp (net) at 5,250rpm, which was less than its C Series replacement, which produced 148bhp (net) at 5,250rpm. A few were manufactured in 1968 with an alloy block for use in the sports-racing MGC GTS.

As the 29AA, the weighty 'six' also powered the shortlived Austin 3-litre saloon, but it differed in a number of respects from the 29G version, most significantly in having its oil pump and attendant sump at the front of the unit rather than at the rear.

Whatever happened to the **MGD**? There has never been such a model although in the 1950s, the MG design register lists EX 188 as 'New Midget- A30 Engine M.G.D', but before the MGB or MGC existed! It was probably never applied to a production car because Jaguar had already made the letter world-famous on its Le Mans-winning D-type.

Although an **MGE** was never built, according to the MG experimental register there were thoughts about so naming the four-seater version of the MGA *qv*, an idea that never came to fruition.

When the ***MGF*** entered volume production at Rover's Longbridge factory on August 4, 1995, it was the first purpose-designed MG sports car to appear since the MGB in 1962.

An MG*F* shows its versatility at the model's second birthday party in 1997.

The **MG*F*** is so rendered because the marque's Rover Group *qv* owner wanted to underline MG's status as a separate make. The decision was made just prior to the model's 1995 announcement, after a preliminary brochure had been printed which refers to the model as the MGF!

The **MG*F*** Cup, administered by the British Racing Drivers' Club, appeared alongside major British Formula Three events for the first time in 1998. Initiated by Rover's MG Cars *qv* subsidiary, it has evolved from MG motor sport activities in Japan and France in 1995 and 1996. Drivers compete in 30 race-prepared MG*F*s tuned to develop close on 200bhp. The total prize fund is £90,000 and includes a new MG*F* VVC as the prize for the overall champion.

The inspiration for the **MG*F* filler cap** came from the Yamaha motorcycle. Gerry McGovern, who was responsible for it and the *F*'s distinctive lines, recalls a lady owner telling him that she loved the petrol filler cap so much she bought the car. I said: "You should have told me that before. I could have probably got you a petrol filler cap and you wouldn't have wasted your money on the rest of the car!"

Based on British Motor Heritage's *qv* revived MGB bodyshell, the 3,946cc V8-engined **MG RV8** roadster was unveiled on October 24, to celebrate the MGB's 30th anniversary, at the 1992 British Motor Show at Birmingham's National Exhibition Centre. Chassis numbers ran from VIN no SARRAWBMBMG000251 to 002233, making a total of 1,983 built.

The 1,098cc **MG 1100**, a derivative of BMC's successful front-wheel-drive Morris 1100, coded ADO 16, was produced between 1962 and 1968 and 116,827 were built. Chassis numbers ran from G/A2S3 (two-door) and G/AS3 (four-door) 101 to 127305.

The 1,275cc **MG 1300** that evolved from the 1100 was far less popular and was built between 1967 and 1971, with 26,240 manufactured. Chassis numbers ran from G/A2S4 118878 (two-door) and G/AS4 (four-door) 118966. A Mark II version followed in 1968 and the final chassis number was G/A2S5 155527.

The **MG 1300** was briefly manufactured in glassfibre-bodied form in Arica, Chile, South America. The plant already produced the Mini-Cooper with a glassfibre body, and the similarly executed 1300 which followed was only slighter longer and heavier than the original. The facility survived for a relatively short time, having been closed down by the Alende government that came to power in the autumn of 1970.

The **Midget** line came to an end with the demise of the TF in 1955, but the name was revived in 1961 with the pricier MG version, coded ADO 47, of the 948cc Austin-Healey Sprite *qv*. A total of 16,080 examples were built between March 1961 and October 1962 and ran

The Midget, as it appeared in 1961. This picture was embargoed until June 30 of that year!

from chassis numbers G/AN 101 to 16183. With the arrival of the 1,098cc engine, the numbers ran from G/AN2 16184 to 25787 in March 1964 and 9,601 were built. The Mark II Midget was produced between March 1964 and October 1966 from G/AN3 25788 to 52411, a total of 22,601 being built.

The 1,275cc Mark III was manufactured between October 1966 and 1969 and the numbers ran from G/AN4 52412 to G/AN5 153920, and 100,345 were built. The 1,493cc Mark III was built between October 1974 and December 1997, from G/AN6 154101 to 229526, and amounted to 73,899 cars.

Kenneth (Ken) Henry Miles (1918-1966), an expatriate Englishman, upheld MG honours both in America and Europe. Born in Sutton Coldfield, Miles raced motorcycles before the war and a Frazer Nash after it. He migrated to Southern California in 1952 and appeared at Pebble Beach in 1953 in a tubular-chassis MG special MGR 1.

Miles had intended to fit a TD unit, but the factory supplied him with a Mark II TD engine that could be bored out to 1,499cc, which endowed the car with a top speed of 112mph. He won, first time out, at Pebble Beach and defeated three OSCAs in the process.

Running in the 1.5-litre class, Miles won nine out of 10 races he entered in 1953. In 1954 he ran a second special MGR 2, the famous 'flying shingle', with low aerodynamic bodywork. The engine was a spare 1,466cc unit that MG had taken to Utah for EX 179's runs in 1954, in which Miles had shared the driving with George Eyston *qv*.

Later in 1955 he shared the wheel of an EX 182 *qv* with Johnnie Lockett at Le Mans and, coming in 12th, was the highest placed MG driver. He and Lockett drove EX 179 again in 1956, but that year Miles gave up driving for MG and switched to von Neumann's Porsches.

Ten years later, in 1966, he and Denny Hulme were placed second at Le Mans in a Ford GT40, but sadly Ken Miles died in August of that year, aged 47, at the Riverside circuit, California, whilst testing a Ford prototype.

The **millionth MG**, built in October 1975, was a unique left-hand-drive roadster finished in Brooklands Green, but with Jubilee GT *qv* livery and wheels. It was exported to America and offered as a prize in a rally organized by the New England MG T Register and the MG Car Club in the United States. The winners were Bryan Wladis and George Cookson. It still survives in the USA and answers to the name of Millie.

The first mechanical passion of MG's chief chassis engineer **Terry Mitchell** (born 1921) was railways, and although unable to obtain a hoped for job at the Great Western Railway's Swindon works he became a clerk at Maidenhead Station. Fortunately, in 1950 he was successfully interviewed by Gerald Palmer *qv* for a job at MG's Cowley drawing office and moved to Abingdon in 1954 when design responsibility was transferred there. He was largely responsible for the EX 179 record-breaker *qv* and its EX 181 *qv* successor. An enthusiastic advocate of the de Dion rear axle, the latter was so equipped. He was thereafter closely involved in the MGA and MGB and, as Roy Brocklehurst's deputy, oversaw the building of MG's first experimental MGB GT V8.

He is also remembered at Abingdon as a prolific special builder; there were no less than 10 of them, of which the most fearsome was an MGA Twin Cam-engined Austin A90! He also essayed his own band saw and lathe, and still maintains a passion for model steam engines.

The 1,994cc MG version of BL's **Montego** saloon, designated LM11, was available between April 1984 and November 1991. Chassis numbers ran from SAXXEYLU7AM100142 to 6255863 and 38,013 were produced.

The 1,994cc MG **Montego Turbo** was available between April 1985 and November 1991. Chassis numbers ran from 172219* to 616700 and 7,355 were built. **Early number.*

Morris' Bodies Branch supplied MG with its bodies from 1935 until facility was closed in 1971. This company had its origins in a coach and carriage business established in 1812 in the city's Quinton Road, prior to the building of the now adjacent railway line. In 1876 it was bought by Edward Henry Hollick, and Messrs Hollick was responsible for bodying, in 1888, J K Starley's electrically-propelled tricycle, probably the first motor vehicle to be made in the city of Coventry.

Hollick's son-in-law, Lancelot W Pratt, subsequently became involved and the firm became Hollick and Pratt. Its first contact with Morris came during the First World War when, in 1915, it began to supply coachwork for his Continental-engined Cowley. In the postwar years, Hollick and Pratt became the principal supplier of Morris bodies, although in 1922 the business had the misfortune to suffer a serious fire. Lancelot Pratt suggested that, as his largest customer, Morris should buy the company. This he did and, on January 1, 1923, the firm was renamed Morris Bodies. Pratt became deputy managing director of W R M Motors in 1923, but he died suddenly during the following year.

When Morris Motors was floated as a public company in 1926, those businesses personally owned by Morris were bought by the newly named firm and became Bodies Branch.

Until 1935, MG's coachbuilding requirements had been provided by the Coventry firm of Carbodies *qv*, but with Abingdon's takeover by Morris Motors, Kimber had to switch his requirements to the in-house facility. The SA saloon and the TA of 1936 were the first MG bodies produced at Quinton Road, which later became responsible for the VA and WA saloons, completed the Y-type, and TC, TD and TF's two-seaters. It also produced the MGA bodies and assembled panels produced by Pressed Steel's Stratton St Margaret plant for the MGB. They were sprayed in a new paint plant and trimmed at the Coventry works prior to dispatch, six at a time, to Abingdon.

In 1966, Bodies Branch became, along with Pressed Steel *qv*, Fisher and Ludlow and Nuffield Metal Products *qv*, Pressed Steel Fisher. This arrangement continued until 1971, when the factory closed following British Leyland's decision to discontinue the Morris Minor for, although its saloon bodies were produced by Nuffield Metal Products, the timber-framed Traveller version had been made in Coventry. When this occurred Pressed Steel became wholly responsible for the MGB.

Morris' Engines Branch in Coventry supplied MG with its engines for most of its existence, apart from the period between 1928 and 1935, when the overhead-camshaft units were manufactured by Wolseley. The business had its origins in the factory established in 1915 by the French armaments manufacturer Hotchkiss et Cie, at Gosford Street, Coventry, when it feared its works at St Denis was likely to be overrun by the Germans. After the war, it approached William Morris and the firm manufactured a copy of the Detroit-built U-type Continental Red Seal engine he had hitherto imported from America. This duly powered the Cowley and better-equipped Oxford that were destined to be the best-selling British cars of the 1920s.

Following the creation in 1923 of Morris Bodies *qv*, Hotchkiss suggested to Morris that he also take over its business. Its problem was that it was unable to keep up with demands for engine production and was being stretched to manufacture 300 units a week.

This was accomplished in 1923, the firm was renamed Morris Engines, and £300,000 was spent on refurbishment at the instigation of the talented Frank Woollard *qv*, formerly of E G Wrigley, who became general manager. Such was the sophistication of Woollard's initiative that by the end of 1924 output had soared to no less than

1,200 engines a week.

When Morris Motors was floated as a public company in 1926, the business was renamed Engines Branch. With the emphasis on growth, there was no room for expansion at Engines' city-centre Gosford Street location, so in 1927 Morris acquired a new 45-acre site in the north of Coventry, at Courthouse Green. In 1929, the foundry that had been established at Cowley in 1919 to produce castings for Hotchkiss was moved to the embryo plant at Bell Green Road. It produced its first castings on the day that William Morris was created a baronet in 1929. Subsequently, one of Britain's most up-to-date car factories of the 1930s was built there, and in 1938 Engines Branch moved to the new premises.

The old Gosford Street factory became the home of Nuffield Mechanisations, and the Bofors gun was manufactured during hostilities. After the war, the premises was taken over by Coventry City Council, and is now part of Coventry University.

In the meantime, in 1927, Morris had taken over Wolseley *qv* and until 1935 it supplied MG with the four and six-cylinder overhead-camshaft engines it required from Birmingham. But in 1935, MG and Wolseley were forced to adopt cheaper pushrod engines, which were designed and built in Coventry. For MG this meant the MPJG and XPAG (both *qv*) units. The latter in its final 1,466cc XPEG form served Abingdon until 1955.

When Riley *qv* production moved from its Coventry location to Abingdon in 1949, the Courthouse Green works became No 1 Factory and Riley's Durbar Avenue plant No 2. It continued to produce the distinctive twin-camshaft/short-pushrod Riley engines that were then dispatched to MG, but when these units ceased production in 1957 the factory became BMC's engine and gearbox reconditioning facility.

The creation of BMC in 1952 had seen the Morris power units phased out and replaced by more up-to-date Longbridge-built A and B Series units. By contrast, Morris Engines was responsible for the design and manufacture of the corporate 2.6-litre C Series *qv* six-cylinder engine of 1954.

Coventry also developed and manufactured the B Series-based 1,588cc Twin Cam engine used in the MGA. Conceived by Gerald Palmer *qv* at Cowley in 1954, it was then handed over to Eddie Maher *qv* at Engines Branch for development, and entered production in 1958.

Courthouse Green closed in 1981, following Sir Michael Edwardes' *qv* downsizing of what had become BL Cars, although Durbar Avenue was taken over by Unipart, which continues its reconditioning activities there.

The Morris Garages was the business from which MG took its name. In 1902, William Morris *qv* had rented former livery stables on the corner of Holywell and Longwall Streets, Oxford, to produce bicycles and motorcycles. These backed on to the old city wall that encircles New College gardens and were soon being used for car repairs and garaging. Sales grew, and in 1907 Morris began to redevelop the area. An impressive new building with a neo-Georgian facade was opened in Longwall (q.v) that in 1910 was named The Morris Garage. However, the demands on space were such that additional premises had to be secured in nearby St Cross Road.

The business became The Morris Garages (W R Morris, Proprietor) in 1913 when showrooms were opened at 36 and 37 Queen Street. The first manager, Frank Barton, was reputed to have been able to sell sand in the Sahara! Further expansion came in 1914, when the large garage in the Clarendon Hotel, in Cornmarket, was acquired.

Not only did the company sell Morris' first car, the Cowley-assembled Oxford of 1913 - the first example had been built at Longwall - but the business also repaired a wide variety of motor vehicles and motorcycles and rented cars for hire.

Yet more sites were occupied in the immediate postwar years, which saw the firm take over premises in Magdalen Street and, significantly for the MG story, a mews garage in Alfred Lane *qv*. Later premises were acquired in George Street and Merton Street. With

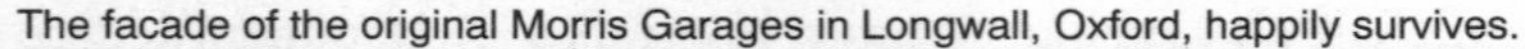

The facade of the original Morris Garages in Longwall, Oxford, happily survives.

One of the many outposts of The Morris Garages in Oxford.

business buoyant, in 1923 the Clarendon garage was rebuilt and modernized.

Barton left in 1919, and his replacement as manager, in March of that year, was Morris' friend Edward Armstead, who in 1908 had bought the rights to a motorcycle that William had produced. Cecil Kimber *qv* joined the business in 1921 as sales manager, but in March 1922 Armstead suddenly resigned and committed suicide a few weeks later. Kimber took over his job and was based at The Garages' Queen Street head office. He was managing a successful business, and in the 14 years between 1911 and 1925 profits amounted to some £50,000.

Such MG stalwarts as Cecil Cousins *qv*, Syd Enever *qv* and Frank Tayler *qv* began their careers with The Morris Garages . . .

The Morris Garages Ltd was incorporated on July 2, 1927, and Cecil Kimber continued as general manager until the creation of The MG Car Company *qv* when, on July 17, 1930, it was announced that Edward Joseph Tobin, the assistant manager, was to take over. This is what Morris' biographers call 'a consolation prize' for the loss to the Garages of its MG car manufacturing arm.

Born in Bangalore, India, and educated in Britain, Tobin had joined the company as a driver in 1906 and was involved in the creation of the original Morris Oxford. After war service in the RFC and RAF, when he was awarded the MSM, he had also been party to the birth of MG.

The Garages had long been in need of new headquarters, having lost the Clarendon garage when the site was redeveloped. Morris bought some old property at the end of St Aldate's, which was demolished and an impressive new showroom and workshop built at the cost of £80,000.

Opened in September 1932, everything had been thought of. *The Morris Owner* noted that there would be "changing rooms for car owners. Visiting motorists will be able to change into their evening clothes when attending functions in the city, without having to trouble about a hotel." At this point the Garages' Queen Street showrooms and George Street offices were closed, along with various other premises scattered throughout the city, although the original Longwall garage remained.

When those businesses personally owned by Lord Nuffield were bought by Morris Motors in 1935, he retained ownership of The Morris Garages, and was also its chairman. At this point Edward Tobin was appointed managing director, a position he held until the late 1950s.

Towards the end of the war, on October 2, 1944, the firm took over the Oxford coachbuilding business of Charles Raworth and Sons *qv*, which had produced the original Morris Oxford bodies.

The Morris Garages, like Morris Motors, was doing well, and this buoyancy was maintained in the years that immediately followed the creation of BMC in 1952. Later, in the 1960s, a new branch opened at Park End Street, Oxford. By this time the ownership of The Garages had passed, in 1961, from its founder to the Viscount Nuffield Auxiliary Trust.

But as the fortunes of the British motor industry began to wain, so did those of The Morris Garages, and late in 1975 it moved from St Aldate's, its home for 43 years, to the outskirts of the city, and a garage in the shadow of the former Pressed Steel factory on the Watlington Road, Cowley, near to its already established commercial vehicle depot. The city centre premises were taken over by the Post Office, and now serve as Oxfordshire's County Court.

The Oxford Times spoke in 1975 of The Garages' plans for expansion and improvement "which will take it forward towards its centenary". But four years later, on December 14, 1979, the firm closed its doors for the last time in the wake of losing its Austin Morris franchise.

Over 100 employees were made redundant, and one of their number, Arthur Exell, later recalled a meeting called by the management on November 6. "Everyone was there, mechanics, labourers, apprentices, typists, clerks, foremen, salesmen. When its

The Morris Garages headquarters, from 1932 until 1975, in St Aldate's, Oxford. It now serves as the Oxford County Court.

spokesman, a Mr Finch, said: 'The Morris Garages are to close', there was a terrible hush over all the employees . . . some were shedding tears, the blow was unbelievable, not a question was asked, and it was Lord Nuffield's pride and joy. He always said 'Morris Garages will always be there' ".

So, after 89 years, the business that gave its name to the MG marque was no more, and today the premises is occupied by Hartwells of Oxford, which sells Ford cars, the very make William Morris intended to challenge when he introduced the Morris Oxford in 1913.

The Morris Garages Chummy, announced in October 1922, was Cecil Kimber's first special-bodied car and a joint venture with Messrs Buist, of Newcastle-upon-Tyne, and the Coventry-based Parkside Garage. Produced very briefly at Longwall *qv* and subsequently Alfred Lane, *qv* on the Cowley and Oxford chassis, the Carbodies *qv* open pastel blue coachwork permitted the back seat passengers, either one adult or two children, who sat opposite one another, to be protected from the elements by the hood when it rained. This contrasted with the more conventional two-seater with dickey seat whose luckless rear occupants were isolated from the raised hood. Its price of 255 guineas (£267.75) was the same as for the standard four-seater standard Cowley.

The Chummy's success prompted Morris, in September 1923, to

offer his own version which, selling for £215, undercut Kimber's effort, production of which ceased that month. A total of 109 Chummies were built between February and September 1923, of which 85 were fitted to the Cowley chassis and 24 to the Oxford frame.

Effectively the first MG, although not badged as such, and described in 1923 by **The Morris Garages** as a Special Morris Cowley, Cecil Kimber had ordered six open two-seater bodies from Oxford coachbuilders Charles Raworth *qv*. Based on the 11.9hp 1,548cc Bullnose, it was available from mid-1923 until late 1924.

The **Morris Minor**, designed by Alec Issigonis *qv*, could have become an MG. When the Nuffield Organisation's managing director, Sir Miles Thomas *qv*, was trying to get the car into production to save the project he envisaged the creation of a two-door 1.1-litre MG version. But the proposal was vetoed by Lord Nuffield in his attempt to kill the entire project. In fact the car entered production as a Morris at Cowley in 1948 and became the first British car to sell a million.

But a version of the Minor was built at Abingdon in 1960 when the American sports car market went into recession and the compound was full of thousands of unsold cars. MG therefore took over assembly of its Traveller estate car and LCV (Light Commercial Vehicle) variants, which were built on the Midget production line. This continued until 1964, by which time the MGB was in strong demand and the operation ceased. Abingdon had produced 10,818 Morris Minor Travellers, 9,147 vans and 49 Pick-ups.

The publicly quoted **Morris Motors,** incorporated in 1926, which purchased the MG Car Company *qv* in 1935, acquired the privately funded W R M Motors, established by William Morris in 1912. Despite the creation of BMC in 1952, it continued to survive as a separate entity until the establishment of the British Leyland Motor Corporation in 1968. It was then briefly replaced by the BMC Morris Division.

The Morris Owner first appeared in May 1924 and initially was the only outlet for MG advertising. It was edited until 1927 by Miles Thomas *qv*, formerly of *The Light Car and Cyclecar*. But it did not carry a great deal of MG news, which no doubt was a factor in Cecil Kimber founding *The MG MaGazine qv* in 1933. The Morris title survived until 1946, when it was renamed, awkwardly, *The New Outlook on Motoring*, which became a more manageable *Motoring* in

1950. This carried MG Car Club *qv* news from 1949 until the arrival of its own *Safety Fast qv* in 1959.

Motor Tramp, by John Heygate (1903-1976), published by Jonathan Cape in 1935, is dedicated 'to CG 1425,' his F-type Magna black tourer with red wings. "Its lines and proportions were so good that one could only tell its size by comparison." The book recounts the author's European travels, twice to Nazi Germany, for which he appears to display some sympathy, although this was beginning to wain during a second visit. Educated at Eton, where his father was a master, and Balliol College, Oxford, he was accompanied on his first trip by the novelist Anthony Powell. Heygate had met him at Evelyn Waugh's London flat and he had wooed and later married, in 1929, Waugh's first wife, although she divorced him in 1936. In 1940 he became Sir John Heygate, having inherited the family baronetage from his uncle.

Without **William Richard Morris** (1877-1963) there could never have been an MG marque. He was well on his way to becoming Britain's largest car maker when, in 1922, Cecil Kimber *qv* was made general manager of The Morris Garages (q.v) in Oxford. Morris achieved this with the famous Bullnose Oxford and Cowley to overtake Ford in 1924 and, apart from the years 1933-34, when his great rival Herbert Austin edged ahead, he maintained this pre-eminent position until the outbreak of the Second World War in 1939.

As Kimber's Oxford-based activities grew, Morris kept a close watch on the growing business and used to visit Edmund Road *qv* every fortnight to examine the accounts.

But in many respects Kimber was fortunate that MG's move to Abingdon coincided with a period when Morris began to spend an increasing amount of time abroad, usually in the Antipodes.

It was this lack of control that allowed Kimber to develop a racing programme at a time when MG production was falling in the years between 1931 and 1935, although industry output was on an upturn.

Miles Thomas *qv* has recorded that Kimber's "intrepidity was not always smiled upon from the chairman's office . . . One of his favourite sayings was 'I challenge you to show me a motor firm that has supported a racing programme and has not had the receiver in' - then he would reel off a long list of names like Sunbeam, Talbot, Vauxhall, Humber, Hillman and others."*

So why did Morris allow Kimber to produce racing cars in the 1930-35 era? The latter maintained in a 1934 paper that his chairman "has always been big enough – and that is his finest characteristic – to

William Morris
Lord Nuffield

let the author have his way".

This was probably because the potent MG Midgets were, in the main, successfully challenging the Austin Seven's hold in 750cc racing and record-breaking. Herbert Austin's discomfiture perhaps took precedence over the limitations of Kimber's fiscal housekeeping.

It was particularly sad that, over the years, Morris fell out with many of his talented managers and engineering colleagues who had contributed so much to the success of Morris Motors. Cecil Kimber's dismissal in November 1941 followed that of Frank Woollard *qv*, who had made Morris Engines a world leader in mass-production techniques, Arthur Rowse, the brilliant Whitworth Scholar, who had transformed Cowley's manufacturing processes, and Leonard Lord *qv*, who revitalized his business in the mid-1930s.

During his lifetime, Morris, who became a Baronet in 1929, a Baron in 1934 and Viscount Nuffield in 1938, donated some £24 million to charitable causes, yet today his name is little known outside motoring circles. The last Morris car was built in 1982, and it is its MG offspring – just one of his handful of British makes – that will survive into the 21st century.
**Lord Austin - the man*, Z E Lambert and R J Wyatt

Motor Panels: see Mayflower Vehicle Systems.

The designation **MPJG** was applied to the 1,292cc four-cylinder pushrod engine used in the TA of 1936. Its 63 x 102mm bore and stroke were inherited from the side-valve Morris Ten of 1932 and subsequently applied to the Series II version of 1935. Following Leonard Lord's rationalization of Morris Motors, it acquired an overhead-valve cylinder head that, in addition to its use in the TA, was also extended to the short-lived Series II Wolseley 10/40 of 1936-37. It was subsequently extended to the Series III Morris Ten, which was only produced in 1938. Used in the TA until it was discontinued in 1939, its cork-lined clutch, which ran in oil, was inherited from the Hotchkiss-manufactured Continental engine that powered the 1919 Bullnose Morrises!

N

Last of the Magnette range, the 1,271cc six-cylinder **N-type** was produced in four variants. The NA, made between March 1934 and November 1936, was available in two, four and occasional four-seater and Airline coupes forms. The NB was the improved version for 1936 and the ND a two-seater. Total production amounted to 738 cars, with chassis numbers from NA0251 and NA0995. The sports-racing NE, built in August and September 1934, ran from chassis numbers NA0516 to NA0522 and accounted for seven cars.

Lord Nuffield: see William Morris.

The Birmingham-based **Nuffield Metal Products** was responsible for the essentials of the Y-type's saloon bodywork. Set up as Morris Motors Pressings Branch in 1939 on land owned by Wolseley *qv*, it was intended to reduce the amount of such work placed outside the business, most significantly to Pressed Steel *qv*. Renamed as above in 1945, it produced the bodies for the Morris Eight Series E, from which the Y-type was derived, that was completed by Morris' Coventry-based Bodies Branch *qv*. Renamed Pressed Steel Fisher in 1966, it survived until 1971 when its mainstay, the Morris Minor *qv*, was discontinued, as did its nearby Coventry counterpart.

MG was part of **The Nuffield Organisation** from 1940 until 1968. It was the name given by Miles Thomas *qv* on his appointment, in May 1940, as vice-chairman and managing director of the group of car, commercial vehicle and component companies established and acquired by William Morris, later Lord Nuffield. Despite the merging of the Austin and Morris businesses to form the British Motor Corporation in 1952, the Nuffield Organisation continued to exist within BMC. It was not until 1968 and the takeover by Leyland that the name finally made a corporate departure.

O

Paradoxically, the **O Series engine** that powered the MG versions of the Montego and the 1984-91 Maestro was conceived with the MGB in mind, although no production version was ever fitted with it.

The need for a replacement was driven by the American emissions regulations, which required that power units could be driven for 50,000 miles without the need for routine maintenance. A major limitation of the B Series engine *qv* was that the distributor skew gear used to wear, with a resulting effect on the spark that in turn affected the exhaust quality of the engine.

Various attempts were made to resolve the problem, but eventually the British Leyland management recognized that the only answer was to redesign the B. It could thus be used in the MGB, as the corporation's largest American seller, and also in other Leyland cars.

But a design constraint was that new machinery had recently been installed for the manufacture of the B Series' crankshaft, so the O Series had to retain this feature, although its nose was lengthened. The B's cast-iron block, in essence, also remained.

Concept approval for the project was given late in 1972 and experimental engines were running in 1973. Created at Longbridge *qv* under Geoffrey Johnson's direction, the O Series was produced in two over-square capacities which retained the same 84mm bore with a choice of strokes: 75mm for the 1,695cc and 89mm for the 1,993cc version.

The belt-driven single overhead camshaft, located in a new aluminium cylinder head, had its distributor prominently driven at right-angles to the shaft.

The engine was soon being experimentally appraised in the MGB, but never in the production version, and from 1978 onwards it was being progressively used throughout the BL range, beginning with the Princess, and soon being followed by the Morris Marina. This left the MG to soldier on with the increasingly archaic B Series unit so that the BL management could then claim that it was the only car to be fitted with it.

The O Series was subsequently fitted to the Princess' heavily revised Ambassador hatchback replacement of 1982. By then it had developed a reputation as a reliable, although unrefined unit.

For its application in the top of the range Montego saloon of 1984, and its MG variant, the engine underwent some revision. As used in the Ambassador it retained the Issigonis-style gearbox-in-sump layout. But like the Maestro hatchback, the Montego used a cheaper end-on gearbox, and in this instance the 2-litre employed a Honda five-speed unit that, perversely, made its debut in the BL product.

This required a new cylinder block to accommodate it, but this meant that it had to be rotated 180 degrees in the car. In consequence, the valve ports were facing the wrong way, so the aluminium cylinder head was more radically revised to allow the manifolding to face the cars' bulkhead, thus permitting a low bonnet line.

The original alternative valve porting gave way to more efficient paired parts, with the result that the single-carburettor version produced more power than its twin-carburettor equivalent. As used in the 2-litre Montego, it developed 102bhp, but the MG version was offered with a Lucas digital system and Bosch injectors, which had proved their worth on the Rover Vitesse. It produced 115bhp, and although the 9.1:1 compression ratio was unchanged, the MG 2.0 EFi version had special pistons and a close-ratio gearbox. The unit was also extended to the MG Maestro.

For the MG Montego Turbo of 1985, the robust O Series acquired a Garrett AiReseach T3 turbocharger which had the effect of increasing output to 150bhp at 5,100rpm. Changes from the mainstream MG included revised pistons, with fully floating gudgeon pins, sodium-cooled exhaust valves and a compression ratio lowered to 8.5:1. Otherwise the block was unchanged, although changes inevitably were made to the induction and exhaust systems, the work being effected in-house, unlike the Metro Turbo, which was undertaken by Lotus. The turbocharger's maximum pressure was 10psi, and after compression, the air passed through an intercooler mounted in front of the radiator before being passed through an SU-type BL carburettor, in place of the usual fuel injection.

The engine was extended to the limited-production MG Maestro Turbo in 1988. The O Series was discontinued with the demise of the Montego Estate car in 1994, although its essentials survive in Rover's current T Series 'four' of 1992, itself based on the O Series-related twin-cam M16 of 1985!

Cecil Kimber's first probable encounter with an **octagon** in a motor industry context occurred in 1915, when he joined the Sheffield-

Simplex company at Tinsley, Yorkshire. The frontage of this 'model' factory, built in 1905, was dominated by a substantial eight-sided tower that housed some of the firm's offices.

Although MG's famous **octagonal badge** *qv* first appeared on the radiator of the 1928 model year 14/40, the eight-sided figure had already featured on the cars' door tread plates as early as 1925. Not only was the 14/40 Mark IV so badged, its dashboard sprouted the symbol in abundance. Octagonal instrument surrounds reappeared on the J2, and the shape was also extended to control knobs, under the bonnet and even the sidelights.

But once the post-1935 Cowley-designed cars appeared, the octagons disappeared, the instruments of the SA saloon were circular, as were those of the TA. But they returned with a vengeance on the VA which, reputedly, has no less than 36 of them!

Whilst the postwar Y-type saloon had octagonal dials, its TC contemporary did not. The TD similarly used round dials and they were partially retained on the MG Magnette Z Series saloon. The TF had octagonal instruments, but they were banished on the MGA and its MGB successor, although the latter's radio speaker grille was

The Y-type's distinctive octagonal instruments that did not feature on its TC contemporary!

roughly octagonal in shape.

But on the 1982 MG Metro *qv* they featured on the exterior, tailgate and wheel trim centres, as well as the facia, heel mat, facia tray mat and on the engine's ribbed rocker cover. The RV8 restricted the redesigned MG badge to the wheel centres and steering wheel boss. It is discreetly applied to the *MGF's* instruments and is also used in moulded form on the steering wheel boss and facia panel.

So committed was Kimber to MG's **octagon** that some factory clocks were eight-sided and the works typewriters had a special key which typed a miniature version of the car badge. He even commissioned Bluemel Brothers to experimentally produce an octagonal steering wheel. But in practice it was found not to slide through the driver's hands as required, so was not proceeded with.

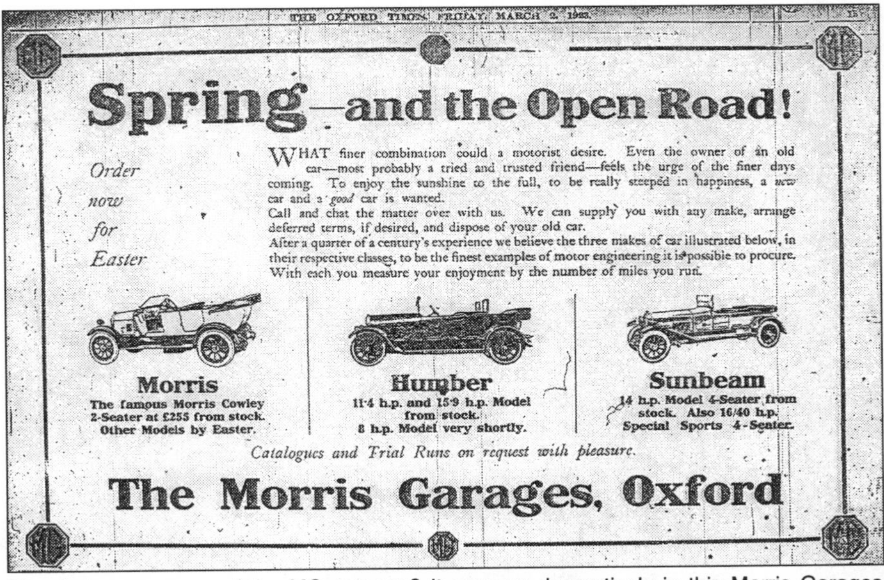

The first appearance of the MG octagon? It appears decoratively in this Morris Garages advertisement in the *Oxford Times* of March 2, 1923.

The **Octagon Car Club** caters for pre-1956 MGs, in other words those cars powered by Morris-based engines up to and including the TF Midget and YB saloon. It was established in 1969 in the Stafford and Stoke area of Britain, with the objective "to arrange social and sporting meetings for the benefit of members, to provide members with information, advice and assistance and generally further the interests in the preservation and renovation of MG cars built prior to 1956".

A magazine, the *Bulletin*, is published every month and regular events such as trials and *concours d'elegance* are held, and in 1972 the club was afforded RAC recognition. An important role is the provision of spares parts for pre-1956 MGs, and the manufacture of hitherto unobtainable spares has been initiated. Details can be obtained the from secretary, Harry Crutchley, at Unit 19, Hollins Business Centre, Rowley Street, Stafford ST16 2RH.

The **octagonally** enhanced striping on the sides and tailgate of the MG Metro were introduced at the suggestion of the MG Car and MG Owners' Clubs. At the instigation of product planning manager Stephen Schlemmer, they were given unique access to the car some months prior to the launch. Schlemmer subsequently became director of Rover Special Products, which was responsible for MG RV8, when the exercise was repeated.

MG's chief body engineer **James (Jim) O'Neill** began his motor industry career in 1936 as an engineering draughtsman with Pressed Steel. He subsequently moved to the adjacent Morris Motors but in 1948 briefly worked for Austin before returning to MG's Cowley drawing office. He was responsible for the drawings of the TD and, when the Abingdon drawing office was reopened in 1954, O'Neill was put in charge of body design. The MGB's distinctive sculptured recesses, subsequently adopted by Don Hayter *qv*, followed a visit to the Geneva show by O'Neill and Syd Enever. He remained at Abingdon until the factory closed its doors in 1980.

Regarded by Cecil Kimber as the first MG, **'Old Number One,'** is the two-seater Morris Garages special that Kimber ran in the 1925 London-to-Exeter trial. Reputedly capable of 80mph, it was powered

"Old Number One" at the MG's factory 50th anniversary celebrations in September 1979. Behind it is the ex-Wilson McComb 18/80.

by a Morris-based CD-type 1,496cc engine enhanced by Hotchkiss with an overhead-valve cylinder head for the Glasgow-built Gilchrist of 1920-23. This was mounted in a modified chassis in which the original Morris three-quarter-elliptic springs were replaced with more suitable half-elliptic ones. Although Morris-based, it therefore bore no resemblance to the production Bullnose, just as the later MGs were essentially unrelated to their Cowley and Birmingham roots.

Graced with a two-seater body by Carbodies *qv*, Kimber and his co-driver Wilfred Mathews completed the event and Kim received a pair of gold cufflinks. Later he gave Mathews a copy of Barré Lyndon's *qv Combat*, which recounts the car's history, dedicating it: "To Wilf - my first passenger in my first MG."

Soon after the event, Kimber sold the car, which had cost £279 or £285 (accounts differ) to build, for £300 to one Harry Turner, of Stockport. It was subsequently sold for £50 to Ronald Davidson in 1930 and auctioned soon afterwards for 11 guineas, later being abandoned to a Manchester scrapyard. Happily, the car was found by an MG employee and, after £12 or £15 had changed hands, this unique 'MG' reverted to corporate ownership.

The May 1933 edition of *The MG MaGazine* reported that MG No 1 . . . "now reposes at the company's works at Abingdon," it having arrived there in the previous year. The article revealed that, in August 1932, "a crack developed across the upper half of the crankcase and lower part of the cylinder block . . . caused by the steering gearbox [sic] breaking away". Fortunately, Barimar was able to effect a repair.

When MG acquired the car it retained its original Oxford registration FC 7900, but after the war this was replaced with a notice that read 'No 1'. Registered for the road in 1950, it acquired the Berkshire number FMO 842, although through the good offices of Wilson McComb *qv*, newly arrived at Abingdon, in 1959 the original FC 7900 was reinstated.

By now, *Old* Number One was being regularly used to publicize MG throughout the world. It was a measure of the importance of the American market that it was displayed at the 1963 New York Motor Show and crossed the Atlantic again in 1975, for a 50th birthday, and in 1988.

Thereafter used by the company for exhibition and promotions, it can now be seen in dignified retirement at the Heritage Motor Centre at Gaydon. Painted red for many years, it has happily now been restored to its original grey livery.

'Old Number One' was produced as a limited-edition porcelain model by British Leyland to inadvertently commemorate MG's 50th anniversary in 1975. A total of 2,500 were made and, with 1,000 being dispatched to America, the 1,500 balance was sold through Austin Morris distributors and dealers. The price was £5 plus VAT.

'Old Speckly Hen' was the name given to the factory's Edmund Road-built 1927 fabric-bodied four-door 14/40 Mark IV saloon, registered WL 3450. It featured one of Kimber's experimental liveries of "a grained tan colour . . . sprinkled with gold dust" and was also available in black. It was one of the variations of the otherwise sober Featherweight Fabric Sports Saloon by Gordon England, of which only 32 examples were built.

After the move to Abingdon, it became a factory hack, but MG's first draughtsman, Keith Smith, lost his job when he borrowed the car for a subversive weekend's motoring. It was broken up in about 1934-35.

The car's unusual colour scheme is said to have been the origin of the name, accorded by the Abingdon workforce, of "awd speckly 'en", which evolved into Old Speckly Hen and so to Old Speckled Hen beer *qv*.

On September 8, 1979, MG celebrated 50 years of its presence in Abingdon, and exactly one week earlier, Morland, the local brewer, had launched **Old Speckled Hen** beer, inspired by Old Speckly Hen, as an anniversary brew. Created at MG's suggestion, no less than 48,000 half-pint bottles of this strong ale, with a 5.2 per cent specific gravity, were produced. In 1990, Morland launched the beer nationally on draught, appropriately at the Boundary House, Abingdon *qv*, the first pint being pulled by Will Corry, chairman of the MG Car Club *qv*. It is currently available in draught, bottled and canned form. Every year the brewery hosts the popular annual Old Speckled Hen run which regularly attracts 100 MGs of all ages.

'One jump ahead' was the name given to an American TV commercial of 1973 that featured the MGB. Its theme was that an MGB on a parachute was dispatched from a cargo plane, whereupon it would float to the ground and then be driven off. But in the first instance the parachute, bearing the MG octagon, failed to open and the car plummeted earthwards. The second attempt was successful and the resulting advertisement was chosen as the outstanding automotive commercial of 1973 by the US Television Commercial Festival.

A mews garage in Alfred Lane, **Oxford**, now named Pusey Lane, was MG's second home and was occupied between February 1922 and

This quiet backwater in Pusey Lane, Oxford, was MG's home between 1922 and 1925 and then a hive of activity. The mews garage was replaced by these flats which, ironically, now proclaim *No Parking.*

September 1925. An outpost of the rambling Morris Garages *qv* empire, and a stone's throw from the busy St Giles thoroughfare, it only measured 20 x 100ft. But such was the growing demand for Kimber's special-bodied Morrises that, during the summer months, work began at 6am and continued until 10 or 12 o'clock at night. When the neighbours complained, Kimber gave orders that work should not begin until 7am! About 25 cars could be squeezed in at a time, but they had to be moved out into the lane before work could start. Later, a lock-up shed in Albion Place, St Ebbes, was acquired to store cars and chassis, but there was some relief when operations were transferred to the new Osberton Radiator factory in nearby Bainton Road. Afterwards the Alfred Lane stables were allocated to Morris Garages' commercial vehicles business and used as a store.

A few years later Alfred Lane was renamed Pusey Lane and the premises subsequently reverted to the motor trade, although the Pusey Street Garage ceased trading in the autumn of 1976, when it was demolished to make way for student flats.

MG's third home in the two years between September 1925 and September 1927 was a new factory, the Osberton Works, in Bainton Road, **Oxford,** so named because it was occupied by in-house Osberton Radiators, which in 1926 was renamed Radiators Branch. It was a business run by Harold Ryder *qv*, who was to take over as MG's managing director following Kimber's dismissal in 1941.

If Barré Lyndon's *qv Combat* is to believed, Kimber was responsible for finding the Bainton Road site, which he passed when on his way from his home in Woodstock Road to his Queen Street offices. It overlooked an old clay pit which had supplied most of northern Oxford with its bricks but had since filled with water. "Kim had discovered the derelict ground . . . and had suggested it held possibilities, as a result of which the radiator works was erected and, as some sort of return for discovering the site, he was given space in the new factory."

But once again demand was such that Kimber's activities required another new home. In his own words: "We first of all started producing in very small amounts in an unwanted bay of the radiator factory; then we required two bays, then six." There was only one thing for it. After a peripatetic five-year existence, MG Cars needed its own purpose-built works and a site was found in Edmund Road, Cowley *qv*, conveniently located near to Morris' factory. Today, the Bainton Road factory still exists as Unipart's Oxford Automotive Components plant.

The original Morris Garage of 1910, located in Longwall Street, **Oxford,** was where Cecil Kimber's Morris Garages Chummy *qv* was briefly produced in 1922, although production was soon transferred, in February 1923, to Alfred Lane *qv*. The most famous of the Bullnose MGs, Old Number One *qv* was built there in 1924-25. Although threatened with demolition in 1979, thankfully the building's facade survives to this day, although the workshops behind have disappeared.

The semi-detached house at 339 Woodstock Road, **Oxford,** was the home in the mid-1920s of Cecil Kimber and his family. It was there that he laid the foundations, with H N Charles *qv*, of the MG marque. Later the Kimbers moved to nearby 1 Hernes Road.

P

Arguably the best of the overhead-camshaft Midgets, the **PA**, built between January 1934 and July 1935, was slightly larger than its J2 predecessor. The 847cc engine benefited from a race-proven three-bearing crankshaft. A total of almost 2,000 were built, of which 1,394 were two-seaters, 498 four-seaters, 28 Airline coupes, 53 chassis supplied to coachbuilders and 27 converted to PBs.

Its **PB** successor, with an enlarged 939cc engine, was the last of the overhead-camshaft MG sports cars. Produced between June 1935 and February 1936, there were 525, of which 408 were two-seaters, 99 four-seaters, 14 Airline coupes and four chassis. Numbers ran from PB0251 to 0776.

Gerald Marley Palmer (born 1911) had overall responsibility for MG car design between 1949 and 1954. An engineer *and* stylist, he was born in Northwood, Middlesex, but his family moved to Southern Rhodesia in 1914, where his father worked as a railway engineer. In 1927 Palmer returned to Britain and gained a degree in mechanical engineering at London University.

In 1928 Gerald Palmer joined Scammell Motors as an apprentice, but he left in 1936 to produce a sports car named the Deroy, after a place in Portugese East Africa, where his father owned a tin mine. A limiting factor was its Scammell Mechanical Horse side-valve engine. On completion, the 27-year-old engineer drove it first to Aston Martin and then to MG at Abingdon, where it was inspected by Cecil Kimber *qv*. Impressed, but unable to offer its designer a job there, MG design having been transferred to Cowley, he nevertheless suggested that Palmer apply for a job there.

After an interview with A V Oak, Gerald Palmer joined Morris Motors in 1937 as a design draughtsman. As such he became involved in the rear end styling of what was provisionally called the MG Ten saloon, which emerged as the Y-type of 1947.

But in 1942 he left Cowley and joined Jowett as chief designer, where he essayed the respected Javelin saloon that entered production in 1947. Sadly, Jowett was over-extended by developing such a sophisticated design, and Palmer returned to Morris Motors in 1949.

Once there he was placed in charge of MG, Riley and Wolseley design and essayed the Wolseley 4/44, MG Magnette and Riley Pathfinder saloons. He was also responsible for the twin-overhead-camshaft version of the BMC B Series engine that was then handed over to Morris Engines' Eddie Maher *qv* and emerged in the MGA Twin Cam of 1958. A similarly enhanced version of BMC's six-cylinder C Series unit was, by contrast, stillborn.

In 1953 he designed an MG Midget sports car with three optional body variations, a competition version with cycle wings, a more traditional rendering, and one with full-width bodywork. This was around the time that Syd Enever was creating EX 175 *qv* and, in the event, it was the latter that entered production in 1955 as the MGA.

Appointed technical director of BMC in 1954, at the end of 1955 Gerald Palmer got his 'marching orders' when his one-time colleague Alec Issigonis *qv* returned to the Corporation. Palmer joined Vauxhall as assistant chief engineer of passenger cars, where he was responsible for the Victor FB's front suspension, and he remained there until his 1972 retirement.

Despite working at Luton, he continued to live at Orchard House, in Tree Lane, Iffley, that he had designed in 1950.

The **Pavlova Leather Company Ltd,** which built what became A Block of the MG factory, was established in Abingdon in 1912. Leather had been a local industry since the 17th century, there being numerous sheep in the Cotswolds and the Berkshire woods provided plenty of oak trees used in the tanning process.

It was in 1835 that W B Bailey opened a business as a fellmonger, who is a dealer in hides and skins, which he then made into parchment. For this he required a plentiful supply of water, and his Spring Grove Works was located at Larkhill Stream, close to what became Cemetery Road, Abingdon.

Bailey lived in what is now Kimber House, occupied by the MG Car Club *qv*, but in 1912 William Batten Bailey, the founder's son, sold the business to London-based Robert Fraser, who, as a patron of the ballet, was a friend of the Russian ballerina Anna Pavlova, and he renamed his purchase The Pavlova Leather Company in her honour. Bailey continued to run it, and clothing, shoe and glove leather were produced which was then supplied to the appropriate manufacturer.

Two years later, in 1914, the First World War broke out and with it came a rapid expansion of Fraser's business to the extent that a second factory was built by A Cox and Sons Ltd, a local builder, to the south

The Pavlova Leather Company's substantial Abingdon factory was demolished in 1998.

of Pavlova's by then much expanded works. Here manufacture, as opposed to leather production, was undertaken. The new premises were known to the workforce as 'the Bailey end' and used to make leather coats and jerkins for troops during the conflict, so ensuring that Pavlova became the largest employer of labour in Abingdon.

With the coming of peace, Bailey gambled on the price of wool skins rising and packed the new factory with thousands of them. But prices tumbled and Fraser was forced to sell out to Alfred Booth and Company, its founder's son, Sir Alfred Booth Bt, also being chairman of Cunard. Pavlova by then had reverted to its own premises, and the wartime building, that was gradually emptied of its huge stock of skins, became the home for the MG Car Company *qv* in 1929. It was named The Pavlova Works *qv* because the leather company still owned it.

Following the outbreak of the Second World War, on Cecil Kimber's *qv* initiative, the Wellworthy piston and ring company, which was looking to expand production in the dark days of 1940, took over the running of the Pavlova factory. The first pistons were for the Rolls-Royce Vulture aero engine, followed in 1941 by those for the Merlin, and the Bristol Mercury, Taurus and Centaurus followed. At its height the factory employed 500 workers, a high proportion of whom were women.

Leather production resumed after the war, and in the early 1980s the Booth interest combined with those of Garnar to form Garnar

Booth. In turn it merged, in 1987, with Pittard of Yeovil, Somerset, to create Pittard Garner. The Abingdon tannery became one of its four operating divisions, responsible for leather clothing and chamois production. Some 80 per cent of output was exported but the world recession of the early 1990s resulted in the business making a £600,000 loss in the first six months of 1993. The factory closed, with the loss of 290 jobs, early in 1994 and was demolished in 1998. Within the space of 14 years Abingdon had lost MG and Pavlova, its one-time two principal employers.

Pavlova Works: see Abingdon Works.

An MG Tri-ang **pedal car** was produced by Walter Lines' Lines Brothers' company during the 1930s. Dating from between *circa* 1932-38, it was a very loose interpretation of the J2 in its original cycle-winged state. Appropriately, wire wheels were fitted.

Arthur George Pendrell (1889-1951) was, with Frank Woollard *qv*, responsible for the JA unit used in the 18/80 MG. London-born Pendrell began his career as a woodworker at Shoreditch Technical College, but he switched to metal and trained in London with the Humpris Gear Company. Subsequently he moved to Coventry and variously worked for Rex motorcycles, British Thomson-Houston magnetos and Hotchkiss prior to its takeover by Morris.

He became chief and later experimental engineer and worked in close co-operation with Woollard on engine design, with the latter applying mass-production techniques to the creative process. A keen motorist, he was an early customer for the Morris Garages' MG Super Sports with polished aluminium body.

George Phillips (1914-1993) was one of three drivers – Dick Jacobs *qv* and Ted Lund *qv* were the others – who drove works-prepared MGs in the early postwar years. Known to his friend as 'Phil', he had been a motorcycle despatch rider in Fleet Street and, as its chief photographer, became one of the founding fathers of the weekly motoring magazine *Autosport* (q.v), which appeared in August 1950, the others being Gregor Grant and John Bolster. He outlived both of them and was able to enliven that publication's 40th birthday party.

A robust, press-on type, he had gravitated to a TC MG in 1947 when his 2-litre Aston Martin deposited a conrod and assorted engine components on the North Circular Road . . . He had purchased the MG from University Motors *qv* and it was accordingly registered MG 7185.

Intent of extracting more performance, he requested help from the factory, and with Abingdon's input in the form of Alec Hounslow *qv*, Phillips succeeded in achieving a second in class in the 1947 Brighton Speed Trials.

Encouraged by this success, he decided to dispense with the original body, and in 1948 he commissioned Harry Lester to produce a lighter, updated two-seater. The only trouble was when it was completed George discovered it was too narrow for the Le Mans regulations, so he had it rebodied again by Ted Goodwin. The result was a similar but more curved rendering, with headlamps incorporated into the radiator grille. Phillips was responsible for his own tuning, "despite knowing nothing about engines".

The car performed sufficiently well to encourage George to enter the first postwar Le Mans race held in 1949 and, in doing so, his was the only MG in the 49-strong field. Co-driving with 'Curly' Dryden, he had to withdraw on the 134th lap after, as he recounted in his inimitable way, "the bloody idiot" had allowed their mechanic, Billy Wickens, to travel in the car on the circuit after he had delivered a spare magnato to cure a misfire. The inevitable disqualification followed.

Phillips, along with Dick Jacobs *qv* and Ted Lund *qv*, was asked by John Thornley in 1949 to drive a works TC in the first production car race at Silverstone, when he was placed 21st.

But he returned to Le Mans in 1950 on his own account, and again had the only MG entered. With another co-driver, Eric Winterbottom, not only did he succeed in finishing in 18th position, but was also second in the 1.5-litre class.

After the event he swapped MG 7185 for a roadgoing TC because he thought it would be an easy car to sell. He drove a works TD in the 1950 TT and was placed 19th and second in class. This in turn led to the creation of EX 172 *qv* for the 1951 Le Mans race.

From George's standpoint, the outcome was far from satisfactory when the car dropped a valve, but after the dust had settled he and Thornley buried their differences. The episode, however, marked the end of his competition career, although he remained at *Autosport* for 17 years.

MGs have long been used by **police** forces throughout the country. The first was delivered in 1930, and by the outbreak of war some 400 were in use throughout the country. They included the 18/80, M-type, PB, TA and VA. This tradition was maintained after the war and the T Series, MGA, MGB and MGB GT V8 were all used by the countrywide constabularies.

One of the more bizarre MG liveries was a TC sold in 1948 by

International Motors, of Los Angeles, that was finished in **polka-dots.** It was so enhanced for 'polka-dot queen' Chili Williams, to match her swimming costume, although *The Autocar's* correspondent thought she looked anything but chilly . . .

Technical editor of *The Motor* for 23 years, **Laurence Evelyn Wood Pomeroy** Jnr (1907-1966) was closely involved with the development of the Zoller supercharger used on the Q and R-type racers and was also a very satisfied VA owner. The son of Vauxhall and Daimler managing director Laurence Pomeroy had no technical qualifications, having failed the entrance examination for Cambridge, so he learnt his engineering directly from his famous father.

In 1929 he went to work for one E T White, who held the British manufacturing rights for the German eccentric vane-type Zoller supercharger. He succeeded in convincing Old Etonian and former Rolls-Royce apprentice Michael McEvoy, and his business partner Henry Laird, that they should form a company to manufacture the Zoller, and Michael McEvoy Ltd was duly established in Derby in 1933. As such it was used by MG, with the type Q4 Zoller fitted to the Q, whilst the R employed the R4A. It was also listed for such unlikely vehicles as the Ford Eight, Lanchester Ten and Austin Ten!

In 1936 Pomeroy went to Germany to consult with Faudi Feinbau, in Frankfurt, who held the Zoller patents, on the design of a new smaller unit and he did not return to Britain until 1937, when he joined the staff of *The Motor*. As technical editor, not only did he contribute readable articles on engineering matters, but they invariably contained a valuable European perspective.

On his appointment and having decided on a choice of car, Pomeroy opted for an MG VA Tickford drophead coupe and in 1938 set down his reason for the decision. "First, for many years I had very pleasant relations with the MG company; second, the general layout of the car was in accordance with my needs – moderate size, high cruising speed, a smooth and silent engine, excellent brakes; third, it had a stiff welded steel chassis, which is vital if one is to obtain satisfaction from a drophead body."

An advocate of front-wheel drive from the 1930s, this portly Edwardian figure, complete with monocle and often seen wielding a sliderule, made a significant contribution to the public's understanding of the motor industry and the history of motor racing.

Because MG produced its cars in relatively low volumes, the **Pressed Steel Company** did not become a supplier until 1953 when it was responsible for the monocoque body of the Z Series Magnette *qv* saloon. Established at Cowley in 1926 by William Morris and J Henry

Schroder, along with the Budd Corporation of Philadelphia, USA, it heralded the arrival in Britain of the mass-produced pressed steel body, so replacing the handmade coachbuilding process that had sufficed hitherto. However, in 1930 Morris withdrew his interest in the firm because other car companies were reluctant to place orders for bodywork from a business owned by a rival.

Pressed Steel remained solely Cowley-based until it began to expand in the postwar years, first with a plant at Linwood, Scotland, in 1947, although more significantly from 1958, at Stratton St Margaret, on the outskirts of Swindon, and some 20 miles from MG's Abingdon Works.

Its first job was the MGB roadster body and thereafter it produced shells for the MGB GT, MGC and MGB GT V8. Apart from the V8 model, they were then dispatched to Morris' Bodies Branch *qv* in Coventry for painting and trimming, prior to a return to Abingdon. When that facility closed in 1971 this later work was done at Cowley, which made more geographic and thus economic sense.

In 1965 Pressed Steel, which had grown to become Britain's largest independent producer of motor bodies, was taken over by the British Motor Corporation, and in 1966 all BMC's pressings businesses were renamed Pressed Steel Fisher.

The last MGB body, a roadster, was completed at Swindon on October 2, 1980, the final GT having been manufactured on the previous day. And that, the company imagined, was the end of the MGB.

But thankfully, the original body tooling was not destroyed, and in 1987 David Bishop *qv*, of British Motor Heritage *qv*, exhumed it, which permitted the MGB roadster body to re-enter production in 1988. It was an initiative that paved the way for the MG RV8 of 1992.

By this time the business had been renamed Rover Bodies and Pressings, and had left its Cowley site, its activities now being solely concentrated at Swindon. It was responsible for the RV8's doors, bonnet and bootlid panels, that model being produced at Cowley in a low-build facility established on the site of the Pressed Steel factory.

MG's **production** at Oxford and Abingdon between 1928 and 1939 amounted to 21,817 cars. Some 819 Morris-based models were built in Oxford in the 1924-27 period, making a total of approximately 22,636 MGs manufactured between 1924 and 1939.

MG sports car production at Abingdon between 1946 and 1980 amounted to 898,988 cars. Added to the 44,936 MG saloons (a figure that includes 877 YT tourers) produced between 1947 and 1958, this makes a total of 943,924.

Austin-Healey manufacture accounted for a further 180,181 cars,

Riley output stood at 19,834 and there were 20,014 Morris Minor Travellers and Vans. This makes a grand total of 1,163,953 cars assembled at Abingdon in the years between 1945 and 1980, and 965,741 MGs produced there between 1929 and 1980.

A painting by James Dugdale of Cecil Kimber at the wheel of "Old Number One" depicting the car in the 1925 Land's End Trial. He received a first-class award.

Q

The **Q-type** racer was powered by the 746cc Zoller-blown engine. Recognized as being too fast for its chassis, this shortcoming paved the way for the R-type *qv*, to which the Q's engine was transferred. Eight were produced between May and October 1934, and chassis numbers ran from QA0251 to QA0258.

The **Q Series** Morris Engines-designed unit formed the basis of the 2.3-litre overhead-valve 'six' that powered the SA of 1936-39. It originally appeared in 65 x 102mm, 2,062cc side-valve form in the Morris Oxford Six for the 1933 season, and was also used in the following year's model, whilst the Fourteen/Sixteen and Eighteen of 1935 all used the Q in varying capacities, the Eighteen's 69 x 102mm and 2,288cc soon being shared with the SA. This featured an overhead-valve conversion and in this form the unit was also used in the Wolseley Sixteen with its original 65mm bore.

Queen Mary was the name allotted to one of the more intriguing MG might-have-beens. Created in about 1934 and designated EX 150, it was powered by a 3.5-litre Morris Twenty-five six-cylinder overhead-valve engine and a body which concealed an advanced central box-section frame. Suspension was all-independent, with equal-length wishbones. Anti-roll bars of varying thickness were tried. It was cancelled when the MG Car Company was sold in 1935 to Morris Motors.

R

The **R-type** was Hubert Charles' last racing MG and his most ambitious design. Designated EX 147, it was a single-seater, powered by the Q-type's Zoller blown engine, but with an ingenious Y-shaped chassis and, progressively, all-independent suspension by torsion bars. Introduced in April 1935, a mere three months before MG withdrew from motor racing, 10 cars were produced between then and June 1935. Chassis numbers ran from RA0251 to RA0260.

The **R Series** engine that briefly powered the MG Maestro *qv* of 1983-84 was derived from the four-cylinder E Series unit with a chain-driven single overhead camshaft developed by BMC for use in the Austin Maxi of 1969. This boxy hatchback accordingly featured the Issigonis system of mounting the gearbox in the engine's sump, which meant that the carburettor adjoined the bulkhead whilst the distributor was exposed at its forward end.

For the Maestro, a cheaper end-on gearbox, in this instance a Volkswagen unit, was adopted. Like the O Series *qv* application in the Montego saloon, this required a redesigned block and inlet and exhaust manifolds. Freed of this constraint, the R Series, along with the alternative A Plus *qv* engine option, was mounted the other way round and consequently the carburettors were then positioned at the front.

Mechanically the R was very similar to its E Series forebear, but instead of a capacity of 1,485cc, it was 1,598cc, the result of a new crankshaft with an 87mm instead of a 81mm stroke. These were considered to be a better match to the valves, which were always considered as being too large for the 1500 Maxi and too small for the 1750 variant . . .

In its basic single SU carburettor form it developed 81bhp at 5,500rpm. The MG version featured a pair of twin-choke Weber 40 DCNF carburettors and developed 104bhp at 6,000rpm. It proved to be an unhappy union and the engine was replaced in April 1984 by the 1.6-litre S Series unit.

The R was likewise not perpetuated in the 1.6-litre version of the Montego saloon of 1984, which was the source of the much improved S Series 'four'. But the topline 2-litre used the O Series *qv* unit that was exclusively fitted in the MG Montego.

Although MG's prewar **racing** activities between 1930 and 1935 were of relatively short duration, there were few frontline British drivers, who did not compete in an MG at one time or another. As Cecil Kimber put it in 1934, "with regard to racing, our policy has always been for other people to do it for us". This reached its peak in 1933/34 with the internationally competitive K3 Magnette; thereafter drivers invariably switched to the new *monoposto* ERA.

MG drivers of this era included Sir Henry Birkin, Norman Black, Charles Dodson, Kaye Don, Doreen Evans, Kenneth Evans, Goldie Gardner, George Eyston, Eddie Hall, Hugh Hamilton, Earl Howe, Ron Horton, Count 'Johnnie' Lurani, G F A Manby Colegrave, C E C Martin, Tazio Nuvolari, Richard Seaman and Whitney Straight.

Amongst postwar British racing drivers who began their careers with MGs were Roy Salvadori and Duncan Hamilton, both of whom ran R-types. Salvadori bought his from R J Symonds and paid £500 for it in 1946, a goodly sum in those days. Leslie Hawthorn, father of later World Champion Mike, sold Hamilton his R-type which he later hillclimbed. TC drivers included farmer Ron Flockhart, who went on to win Le Mans, and Archie Scott-Brown campaigned a TD. Denny Hulme, World Champion in 1967, began his racing career with a TF 1500, subsequently tracked it down and by 1980 was happily restoring it.

Across the Atlantic, Phil Hill began racing a TC in 1949 and Briggs Cunningham campaigned a supercharged example. Carroll Shelby ran a TC for a time because he believed, erroneously, that it cornered better than the XK120 Jaguar.

But the postwar MG racing story really belongs to that doughty trio of Dick Jacobs, Ted Lund and George Phillips (all *qv*).

The make's greatest international **racing success** came in 1933 when MG won the team prize in that year's Italian Mille Miglia race. Earl Howe led a team of three K3s, and the car driven by George Eyston and Count 'Johnnie' Lurani, that won the 1,100cc class, was K.3003, the same MG in which Nuvolari triumphed in the TT later that year. Lurani later wrote of the *Gran Premio Brescia* award, "an imposing, bronze Roman chariot bigger than our little cars, made it seem a big victory". The first 10 places were taken by Alfa Romeo!

MG's greatest British **racing success** also came in 1933 when Tazio

Nuvolari, in what by then was Whitney Straight's K3 (K.3003), with the factory's Alec Hounslow *qv* as mechanic, won the Tourist Trophy race at Ulster, and its class, at an average speed of 78.65mph. Hugh Hamilton was second in a J4. Appropriately, as it transpired, the first prize of £300, and class award of £300, were donated by Sir William Morris. Hounslow later commented of Nuvolari: "He was as safe as houses . . . a proper marvel."

Cecil Kimber *qv* was responsible for the design of the MG **radiator,** the shape having been sculpted from a block of beech, which was pared down on his instructions by a carpenter from elsewhere in The Morris Garages *qv* business. The honeycomb core was the work of Radiator Branch's Ron Goddard. It first appeared with a distinctive central strip on the 18/80 model of 1928, but in 1931 was offered on the Mark II with optional painted slats, which echoed the colour of the car's upholstery. So the definitive slatted MG radiator appeared, but it was not standardized until the arrival for the 1936 model year of the PB Midget and N-type Magnette. From thereafter it remained a feature of MG sports cars and saloons until, in the former instance, the demise of the TF in 1955 and the ZB Magnette in 1958. The much diminished, although still split rendering first appeared on EX 172 *qv*, the special-bodied TD that George Phillips ran at Le Mans in 1951, and it arrived in 1955 on the production MGA. It has remained a distinctive feature of the marque ever since.

The **rarest** production MG is the sports-racing Mark III *qv* 18/100, or Tigress, announced in the depression year of 1930. Just five examples were produced between 1930 and 1933, selling for £895, making it the most expensive prewar MG. The two survivors, dating from 1930, are registered JB 855 and GH 3501, and were both owned for many years by Christopher Barker, of Winchester, who bought them in 1935 and 1943 respectively.

Of the remaining cars, two—WL 9233 and WL 9285—were sold in 1930, and the third—GS 2211—in 1931. MG was left with 20 Carbodies four-seater touring bodies of the original batch, which were subsequently used on the Mark I 18/80 chassis to create in 1931 the today coveted Speed Model, that was also available with metal bodywork.

The **rarest postwar** MG is the MGA 1600 De Luxe, of which 395 were produced in 1960 to 1962. No mention was made of the model in contemporary publicity material and no price ever quoted. It was based on the superfluous chassis and special running gear, all-round Dunlop disc brakes and knock-off wheels, of the MGA Twin Cam, and like that model it was built in touring and coupes forms. Production began in June 1960 and continued until June 1962. By then 70 tourers and 12 coupes had been built. The Mark II version was the more popular of the two, there being 313 in all, 290 roadsters and 23 closed cars.

Probably the **rarest unofficial** MG was the MGB SEC conversion, of which there were five, produced by Abingdon Classic Cars, based in the town, at about the time of the MGB's demise in the autumn of 1980. It was then that the MG Owners' Club *qv* announced that it was launching its MGB SEC, for Special Edition Classic, at *Thorougbred and Classic Cars*' Classic Car Show, held at Earls Court between October 3 and 5, 1980. The company was established by the hitherto Lower Wooton, Abingdon-based workforce of Mallalieu Cars that transformed rusting Mark VI Bentley saloons into open sports cars. The five B's so converted were three turbocharged four-cylinder cars and two V8s, one left-hand drive and one right-hand drive, before the firm went out of business.

Established in the 19th century, **Charles Raworth and Sons Ltd,** of Oxford, was responsible for bodying Cecil Kimber's second attempt at marketing a sports car in 1923. The links with Morris went back to 1913, when Raworth supplied the bodies for the Oxford model, although it was unable to cope with demand and the contract was taken over by Hollick and Pratt, which evolved into Morris' Bodies Branch *qv*. Cecil Kimber *qv* ordered six open two-seater bodies from Raworth in 1923, which were fitted to the 11.9 Morris Cowley chassis, but the finished product sold for an uncompetitive £350 and the experiment was not repeated. With works at 1-3 Speedwell Street, in the centre of Oxford, the premises incorporated part of the old city wall and stretched to Bridewell Square.

The business survived independently until 1944, when it was taken over by The Morris Garages *qv* and moved to its St Aldate's headquarters.

Like in many firms of the inter-war years, craftsmen worked in primitive conditions and, on the coldest days, the management had to be encouraged to provide coke for the single stove. On vacating Raworth's confined premises, employee Phil Austin recalled that "the boss said 'get some thin paint to make it look a little better'".

Although **red** has long been regarded as a traditional MG colour, of the 2,083 J2s built in 1932-34, it came third with 517 cars so finished, behind black in combination with another colour (585) and green (526). Its identification with the MG was underlined by its use on the

Three Muskeeters *qv* trials cars of the 1937-39 era, all of which had red bodywork.

When the TC entered production in September 1945, it was only available in black, with red, green or beige upholstery. Happily works manager Cec Cousins *qv* believed that this was not appropriate for sports cars and, from about September 1946, the cars were also available in MG Red and Almond Green. He was promptly carpeted by Nuffield's Sir Miles Thomas *qv*, but more colours were soon added.

Riley (Coventry) Ltd was personally purchased by Viscount Nuffield in 1938, an initiative that caused many raised eyebrows at Abingdon. Riley, after all, had been a rival to MG both on the race track, and its well-appointed saloons of the late 1930s were occupying a similar market sector. In a shake-up of Nuffield manufacturing, car production was transferred to the MG factory at Abingdon in May 1949, although incredibly, the original idea was to move MG to Coventry until George Propert *qv* and John Thornley *qv* voiced the strongest possible opposition! Nevertheless, the Riley plant in Durbar Avenue, Coventry, continued to manufacture engines.

Of the 10,504 examples of the 1.5-litre Riley RMA saloon produced between 1946 and 1953, 5,479 were built at Abingdon. It also manufactured all 3,446 examples of its RME derivative of 1952-55 vintage.

This was the last of the 1.5-litre Riley-engined cars, and the One Point Five saloon, introduced in 1957, used the BMC B Series engine. MG built 150 that year before production was transferred to Longbridge.

There were 6,900 examples of the 2.5-litre RMB saloon of 1946, and Abingdon was responsible for 5,639 of them. It also made 386 of the 507 RMC roadsters and produced 498 of the 502 RMD drophead coupes.

The Gerald Palmer *qv*-designed 2.5-litre Pathfinder saloon of 1953-57 vintage accounted for a further 5,152 cars and was replaced by the BMC C Series 2.6-litre-engined Two-Point-Six. This was the last Riley produced at Abingdon, and a mere 434 were built in 1957 and 1958 before assembly was transferred to Cowley, where a further 1,566 were made. The last Riley was manufactured in 1969.

Up until 1975, when a rolling road was installed at Abingdon, all MGs were **road-tested,** a practice that reached back to the firm's prewar origins. The route then took in the villages of Marcham and Frilford, but was abbreviated after the war to avoid the Frilford element. From the late 1940s it was again reduced to eliminate

Marcham, and drivers turned round at the Sheepstead Crossroads. But in 1973-75 the MGB GT V8 was the subject of an extended run to take in the Frilford Crossroads and a length of the A420.

The **Rover Group** is current custodian of the MG name. The name took effect on the occasion of BL plc's annual meeting held on July 7, 1986, and was a reflection of chairman Graham Day's move up-market for the then nationalized company. It returned to the private sector in 1988, when it was bought by British Aerospace *qv*, which in turn sold it, in 1994, to BMW *qv*.

The so-called **'rubber' bumpers** fitted to the MGB and Sprite from the 1975 season in deference to American safety regulations, were only experimentally made of rubber. The resulting units, produced in conjunction with the Cable Belt Company, proved difficult to manufacture, so MG turned to Marley Foam, of Lenham, Kent, which produced the moulded polyurethane bumpers eventually used.

Harold Alfred Ryder (1888-1950) was MG's managing director following the departure in 1941 of Cecil Kimber and remained so until his resignation in 1947. Born in Birmingham and educated there, between 1914 and 1919, Ryder was general manager of Coventry radiator manufacturer Doherty Motor Components, a firm which supplied radiators for the 1913 Morris Oxford. Doherty was unable to cope with Morris' demands, and encouraged Ryder to leave and move to Oxford. Ryder, and his colleague A L Davies, bought a former roller skating rink in Osberton Road, in the city, and he became managing director of the business that was accordingly named Osberton Radiators. It was in turn bought by William Morris *qv* in 1923, but it soon outstripped the confines of the original site and moved to a new additional purpose-built factory in nearby Bainton Road, Oxford *qv*, which between 1925 and 1927 was MG's home.

In 1926 the business was renamed Radiators Branch, Ryder was promoted to general manager, and he also became a Morris director. In 1936 he became one of Morris Motors' three managing directors when Oliver Boden, formerly Wolseley's managing director, was made vice-chairman. With Boden's sudden death in 1940, Miles Thomas *qv* for one believed that Ryder would take over, but Thomas got the job himself.

Although still running the radiator business, after Kimber's departure Ryder was responsible for MG and although not a motor sport enthusiast, he did sanction some activity in 1945. He was sacked by Lord Nuffield in 1947 when he was one of six directors asked to 'resign'. Harold Ryder died suddenly in October 1950 at the age of 62.

Cecil Kimber (centre) with MG racing drivers "Hammy" Hamilton (left) and Earl Howe, who upheld MG laurels on the circuits of Britain and Europe.

S

First of the MGs to have been designed at Cowley, the 2,322cc six-cylinder **SA** sports saloon was also offered in Tickford drophead coupe and four-seater touring forms by Charlesworth. A popular model, a total of 2,738 were built between March 1936 and September 1939. Chassis numbers ran from SA0251 to SA2988.

The **SA** was one of the first MGs to be offered with the option of a radio. A variety of sets were available from the Philco type KT728-T at 18 guineas to Masterradio type 101 at 13.5 guineas. The aerial was three guineas extra.

'Safety Fast', the slogan that MG first used in 1929, was devised by publicity manager George Tuck *qv*. One day in the autumn of that year he was following a bus in Oxford which bore a *Safety First* sign on its rear and displayed a triangle to indicate the fitment of four-wheel brakes. Tuck saw the potential of a Safety Fast derivation and Cecil Kimber immediately responded to his initiative. It also chimed with the public because Stanley Baldwin's 1928 Conservative government had been elected on the *Safety First* catchphrase.

***Safety Fast*,** that took its name from MG's famous slogan, is the MG Car Club's *qv* monthly magazine that first appeared in April 1959 as a factory-backed publication under the editorship of Wilson McComb *qv*. Its arrival came 24 years to the month since the appearance of its predecessor, *The Sports Car qv*, and it survived in this form until 1969, when the Club became an independent organization. It has been published ever since and is currently edited by Paddy Willmer.

MG raised its corporate profile in America, by and away its largest market, with regular appearances at the **Sebring** motor racing circuit. Although never victorious, class victories were regularly achieved and, from 1956 until 1968, the BMC Competitions Department-*qv* prepared MGs for these transatlantic long-distance races, despite initially having withdrawn from such competitions in Britain and Europe.

Prior to 1956, MG cars and Specials were entered by enthusiasts. The first event, at the former B-17 bomber base at Sebring, Florida, complete with Le Mans start, was run on December 31, 1950. Named the Sam Collier Memorial Grand Prix, coincidentally a one-time MG driver, the field included a sprinkling of MGs, and the highest-placed car in the six-hour race was a TC driven by John Van Driel, which was fifth, and MGs won their class.

The second race, extended to 12 hours, was run in March 1952. Van Driel shared the driving with David Ash *qv* and took their MG Special into fourth place. In 1954, the MG-based Motto Special driven by Gus Erhman and Fred Allen was placed 11th. But the make's waining fortunes were reflected by 38th and 39th positions. A new model was clearly needed to replace the T Series Midgets.

BMC entered a team of three MGAs for the 1956 12-hour race, driven by seasoned T Series drivers Kinchloe/Spitler, Ash/Erhman and Allen/Van Driel. All completed the event, were placed 19th, 20th and 22nd, and MG won the team prize.

Later, in 1959, a pair of Abingdon-prepared MGA Twin Cam coupes were placed second and third in class and repeated the achievement in 1960.

With the arrival of the MGB, an example came third in class in 1964, in 1965 one was second, and one of the Dick Jacobs *qv* factory-prepared Midgets, driven by Andrew Hedges and Roger Mac, took a class victory. It coincided with the make's last entries at Le Mans, but MG continued to appear for a time at the Florida circuit.

Further class wins for the MGB came in 1966 and 1967. The lightweight GTS, powered by an aluminium version of the 3-litre MGC engine, won the prototype class at Sebring in 1968, the year in which British Leyland turned the lights out.

Before the **Second World War** MG had a workforce of some 350, which rose to about 1,500, 40 per cent of whom were women. The press shop began making shell racks and a small contract was obtained for overhauling Vickers Armstrong's Carden-Lloyd light tank. It also undertook similar work on the more substantial Matilda. Later Abingdon took over the assembly of Crusader tank, as the principal contractor to in-house Nuffield Mechanisations, which attained an output of 10 a week.

But by far and away the most significant contract was the one Cecil Kimber secured in 1940 for the front section of the Armstrong Whitworth Albemarle bomber, that was specifically designed to be made in sections. Abingdon was responsible for the complex nose portion, whilst Rover built the wings, and furniture manufacturer Harris Lebus the tail section. Eventually MG made 653, and a further 285 were completed because, originally, four other firms were engaged in producing this same G1 unit. But they all dropped out, leaving Abingdon alone in the field, although by then Kimber had been sacked, obstensibly for obtaining the commission.

Then there were Lancaster bomber engine mountings and, prior to D-Day, hundreds of Churchill tanks were converted for use as flailing minesweepers, and thousands of sets of waterproofing equipment were produced. The sensitive work was allocated to MG by Sir Miles Thomas *qv* because he believed the quiet backwater of Abingdon would not be as likely to attract the attention of a technically minded spy . . .

Queen Elizabeth II's **Silver Jubilee** in 1977 was celebrated by MG with the production of a single MGB GT. The car was blue-painted and featured a silver side flash similar to the gold one used on the 1975 Jubilee cars. It contained the logo of the Jubilee Year, and was then raffled in aid of a Vale of the White Horse charity appeal, the local authority area in which Abingdon is now located.

In 1924, The Morris Garages' Alfred Lane, Oxford *qv* premises was used to prepare a six-cylinder Bullnose Morris. Enhanced with handsome MG-style four-seater touring coachwork, it appeared nightly in the play, ***Six Cylinder Love***, starring Edna Best, which was staged in the Oxford Theatre during November.

Sydney V Smith was the Morris director responsible for MG following Harold Ryder's *qv* resignation in 1947. He joined Wolseley in 1912 and returned to the business after war service in 1919. Smith was with Wolseley when Morris bought the business in 1927, and in 1933 he moved to Morris Motors at Cowley. Three years later he became works manager, and manager of the cars branch in 1940.

A major reorganization of manufacturing in 1947 saw Smith promoted to director of assembly facilities, with responsibility for Morris factories in the Oxford region, one of which was MG. Irreverently known at Abingdon as 'Hitler' Smith, on account of a dictatorial moustache, he was subsequently promoted to the post of BMC's engineering co-ordinator. Responsibility for MG's affairs passed to Morris Motors' local director R E T Crouch.

The Sports Car first appeared in April 1935 and replaced *The MG MaGazine*. Unlike its predecessor, it was published monthly and described as "the official journal of the MG Car Company and The

MG Car Club". However, the publisher/editor, F L M Harris *qv*, gave it a more generic title, no doubt to appeal to a wider public and advertisers than just MG owners.

To these ends, *Brooklands Track and Air*, established in 1932 as the *Brooklands Aerodrome Magazine*, was acquired and its editor/proprietor, Wing Commander O V Holmes, became a contributor, as did the assistant editor, a youthful William Boddy, later to become, in 1945, the distinguished editor of *Motor Sport*.

Published from offices at 12 Holborn, London EC1, by F L M Harris Ltd, *The Sports Car's* last issue was October 1939. It then carried a message from Cecil Kimber hoping, if circumstances permitted, there would be a 'war edition'. In the event there was not, and sadly the title died with its editor in 1945.

When, in the late 1950s, John Thornley *qv* wanted to revive the idea of a marque journal, his original thought was to resurrect *The Sports Car* name. Unfortunately, by then there were two magazines with similar names, *Sports Car and Lotus Owner* and *Sports Cars Illustrated*, so he opted instead for MG's own slogan of *Safety Fast qv*.

'The Sports Car America Loved First,' the slogan adopted by British Leyland Motors Inc in 1972, was conceived by its advertising agents, Bozell Jacobs, and thereafter appeared on MG promotional material until the end came in 1980.

Spridget, the name universally applied to the Sprite and Midget, was first used at the Abingdon factory. It was initially used to describe hybrid Sprites fitted with MG Midget grilles, of which there were a number at the plant.

Adrian Morgan Squire (1909-1940) joined MG in 1929 and had left by 1931, but during his time at Abingdon he acquired the reputation of never turning out a bad job. A chassis design draughtsman, he worked with Jack Daniels *qv* and both were inspired by H N Charles *qv*. Squire left to produce his own car, the finance was supplied by Jock Manby Colegrave, and Squire Motors maintained the K3 Magnette (K.3004) he kept until 1935.

By then the Squire, announced in 1934, was a reality and Daniels had executed the chassis drawings in his spare time. But only seven examples of this costly sports car were built at Remenham Hill, near Henley-on-Thames, until 1937. MG influences are readily apparent in its sliding-trunnion suspension and scuttle-mounted oil tank, *à la* C-type, J2-inspired bodywork and preselector gearbox in the K3 Magnette idiom.

After the MG's racing department closed, Alec Hounslow *qv*

worked at Remenham before returning to Abingdon. Squire moved to Lagonda and then to the Bristol Aeroplane Company before being killed in 1940 during a daylight air raid on Filton at the age of 30.

Although MGs have featured on **stamps** throughout the world, the make did not appear on a British one until 1996. As early as 1953 a YB saloon was the subject of a Yugoslavian stamp, whilst in 1974 the Republic of Equatorial Guinea featured an Airline saloon that was inadvertently captioned 'Panhard and Levassor 1934', and that vehicle was identified as an 'MG Midget PA 1934'! A TC featured on one of a set of six stamps issued by Dominica in 1983, and a Guinea-Bissau issue of 1984 included a 1932 MG Midget. An MGA roadster appeared on a 40c Tuvalu stamp and the coupe on a 1984 Madeira one. A $1.50 Grenades of St Vincent issue featured an MGB GT which was one of a 1986 series printed to celebrate the centenary of the motor car.

It was the 100th anniversary of the British motor industry in 1996 that prompted the Royal Mail to issue, on October 1, a series of commemorative stamps. It goes without saying that the 26p one for First Class letters featured an MG TD!

One of the prime movers in the revival of the MG marque, Rover's **Nick Stephenson**, is at the time of writing (1998) the Group's design and engineering director. Educated at Ampleforth College, Nicholas John Stephenson (born 1948) graduated from Queen Mary College, London University, with an engineering degree in 1970. After eight years spent mostly at Perkins Engines, he joined Leyland in 1978 as advanced power-train planning manager and subsequently became, in 1983/84, vehicle planning manager on the Austins Maestro and Montego (LM 10/11).

It was Stephenson's time, in 1991-94, as director of forward planning that saw the emergence of the PR3 concept that subsequently emerged as the MG*F*. A dedicated racing competitor and enthusiast, he pragmatically implemented a unique development programme for the project. Just two aspects were to commission outside agencies to speedily design three experimental cars (PR1, PR2, PR3) and the ultimately fruitful partnership with The Mayflower Corporation *qv*.

Abingdon-born **Henry W Stone** (1910-1990) joined MG in 1930 and initially worked on the assembly line, but in February 1933 he was recruited to the racing department to work after hours. Such was the impression he made on Reg Jackson *qv* that Henry became a full-time member of the 'insomnia crew' of racing mechanics. He was closely involved in its activities until they ceased in mid-1935.

After the war Stone helped to build the prototype TC "from bits" and he then joined his longstanding friend and colleague Alec Hounslow *qv* in the revived experimental department that handled MG's initially low-profile racing activities.

From thereon Henry was involved in every MG competitive initiative from George Phillips' 1951 Le Mans entry to the EX182 Le Mans cars and the EX 179 and 181 record-breakers (all *qv*). He retired in 1974 but, happily, Dick Knudson *qv* set down Stone's memories of Abingdon in the fascinating and well-illustrated *MG Mania: The Insomnia Crew*, published in 1988.

The **stroke** of at least two engines used by MG are far older than might be imagined. The 102mm one of the four and six-cylinder engines used in the TA, SA and VA of the 1935-39 era date back to the unrelated Morris engine used in the Morris Oxford of 1915. And that was the US-designed Continental U type Red Seal unit, later copied for Morris by Hotchkiss.

The 89mm stroke of the BMC B Series *qv* 'four' that was used in the ZA Magnette, MGA and MGB until 1980 dates back to the side-valve Austin Ten unit of 1932!

A young model meets an old one! Stripped deckchair-style seats and a bikini in natural harmony on MGBs for 1977. A revised facia also featured.

T

First of the pushrod-engined Midgets, the 1,292cc **TA** Series was built between March 1936 and April 1939. Of the 3,003 cars produced, the overwhelming majority of 2,722 were open two-seaters, the balance being made up of 260 drophead coupes, two Airline coupes, eight unassembled for export and 11 chassis, of which 10 were dispatched to Australia. Chassis numbers ran from TA0251 to TA3253.

The drophead coupe body Park Ward produced in 1947 for the Mark VI Bentley was inspired by a special-bodied **TA** built in 1945 for the business' founder, Charlie Ward. Based on a secondhand chassis, its performance was enhanced by an Arnott supercharger, and this MG, registered KPB 999, was subsequently modified with a cowled radiator and lengthened front wings.

The TA Series' 1,250cc **TB** successor, produced between April and October 1939, was the first MG to be powered by the Morris Engines Branch-designed XPAG 1.2-litre engine. A total of 379 cars were produced, of which 319 were open two-seaters and the other 60 Tickford *qv* drophead coupes. Chassis numbers ran from TB0251 to TB 0629.

MG's postwar production began in September 1945 with the 1,250cc **TC** Series Midget, which was discontinued in November 1949. There were 10,000 built; right-hand-drive exports accounted for 4,497 cars and 2,001 were dispatched to America. A further 84 were CKD cars for Eire and there were 10 special-bodied cars. Chassis numbers ran from TC0252 to TC10251.

The most famous **TC customer** was the Duke of Edinburgh, who delighted in taking the present Queen for a ride in it. He bought the car (TC1362) in August 1947. The then Princess Elizabeth described the MG, registered HXD 99, to Betty Shrew, a London-based writer who was writing a Royal Wedding book. The Princess told her: "Philip enjoys driving and does it fast! He has his own tiny MG, which he is very proud of . . . it is like sitting on the road and bus wheels are almost higher than one's head!" Prince Philip sold his TC in November 1948.

The **TC** also found plenty of customers amongst Hollywood film stars, mostly supplied by the go-ahead International Motors, of Los Angeles. TC owners included Dick Powell, whose car was fitted with a Marshall-Nordec supercharger, and Yvonne de Carlo had a green example. Producer Billy Wilder was a TC owner, as were Bette Davis and Robert Stack, and the make was also similarly promoted by Gary Cooper, Nat 'King' Cole, Clark Gable and Mel Torme.

Initiated unofficially at Abingdon, the 1,250cc **TD** was numerically the most successful of the T Series Midgets and was responsible for establishing MG on the American market. Produced between November 1949 and August 1953, of the 29,664 cars produced, no less than 80 per cent, or 23,488, were exported to North America.

The **TD** was offered in tuned Mark II form during its 1950 to 1953 production life. Accorded the TD/C, for Competition, chassis prefix, the first example was 1123 and the cars were fitted with the TC Stage I cylinder head in varying stages of tune giving 57 to 61bhp. Of the 1,710 Mark IIs, 1,593 were exported to America.

No less than 166 **TDs**, 157 in left-hand-drive and nine in right-hand-drive form, were bodied by coachbuilders invariably based in continental Europe. In Switzerland, for instance, Ghia-Aigle produced at least single examples of a TD-based roadster and coupe.

In Germany, many more were transformed by local coachbuilders for American GIs to take home. Schlomer made an open two-seater, Weidenhausen produced at least one Ferrari-like open two-seater and Wendler of Reutlingen made both TD-based coupes and roadsters.

Some 25 **TD**s were bodied by Fritz Hennefarth, of Bad Cannstatt, Stuttgart, that superficially resembled the Abingdon product. These were supplied by German MG importer J A Woodhouse, of Cologne, and the work was undertaken from photographs!

Initially, Hennefarth was supplied with a single left-hand-drive rolling chassis and it produced an approximate rendering of the standard body using only hand tools. Inevitably there were differences, for example the wing line was somewhat lower than the original, and the scuttle humps less pronounced. German electrics mingled with Lucas components and the seats were also bespoke. After this first car a further two batches of chassis followed.

The firm also specialized in repairing MGs, and some T Series Midgets were fitted with completely new German front ends. Its favourite refurbishment was an MG with a British left-hand side and a completely new German right-hand side! The company also produced the so-called Jaguar MG, with a rather XK120-like body on the TD chassis for MG dealer Christian Odendahl, in Frankfurt.

The **TD** was the inspiration for Siata's Rallye tourer of 1951, from a Turin company that had specialized in special-bodied Fiats. The Rallye was based on the Fiat 1400, but only a handful were built.

The **TD** was also the source for the Lafer MP, which ingeniously concealed a floorpan and engine of the locally built Brazilian Beetle. Initially 1,500cc, but later 1,600cc-powered, it soon gained some popularity in Europe and by 1980 was being sold in 17 European countries, with Italy the best customer.

Last of the T Series cars, the **TF** Midget, built between September 1953 and April 1955, was at first 1,250cc-engined and from late 1954 was produced in 1,466cc form. Chassis numbers ran from HD501 to 10100. Total production was 9,600 of which 6 200 cars were 1,250cc-engined and the 3,400 balance 1,466cc-powered.

Like the TD, the **TF** inspired a number of replicas, of which the Nottingham-built Spartan was one. Introduced in 1973, the Ford-powered glassfibre-bodied car initially used the Triumph Herald or Vitesse chassis, but later Jim McIntrye introduced his own box-section frame.

Of all the cars based on the TF, the **TF 1700**, from longtime T Series parts supplier Alistair Naylor, was the most faithful. Launched at the 1984 British Motor Show, it was produced by Naylor Cars plc, of Pawson Street, Bradford, the first cars being delivered in March 1985. The TF 1700 was powered by Rover's 1.7-litre O Series engine *qv*, but a little over a year later, and after about 100 cars had been completed, the company went into receivership in April 1986 and was taken over by engineer Maurice Hutson. He renamed the business, which remained on the same premises, Naylor Sports Cars Ltd. Still listed at the time of writing (1998), what has become the Hutson Motor Company now also produces parts and panels for the MGB GT and E-type Jaguar.

John (Jack) Tatlow (1902-1965) was MG's general manager from 1949 until 1953. Coventry-born Jack Tatlow, who was educated, like Leonard Lord *qv*, at the city's Bablake School, studied engineering at Coventry Technical College. He joined Rover as a trainee in 1918, but in 1924 moved to Riley *qv* as outside sales representative. He rose to be Riley's service manager and, on Nuffield buying the business in 1938, became works manager and was promoted to general manager in 1940.

When the Nuffield Organisation *qv* transferred Riley car production to Abingdon in 1949, Tatlow became MG's general manager, although he continued to live at his home at Bagington Road, Coventry, and commuted to the MG factory.

Another Riley recruit was Arnold Farrar, who became service manager, replacing John Thornley, who became assistant general manager. Tatlow took over from Cec Cousins *qv*, who had been holding the fort since 1944. He was respected at MG for his fair dealing and was replaced in November 1952 by John Thornley *qv* . In February 1953 Tatlow became general manager of Morris Commercial Cars in Birmingham, and in 1956 he was appointed a local Austin director and before retirement he returned to Morris Commerical as branch manager in 1960.

The dedication in John Thornley's *Maintaining the Breed* is: "To **Frankie Tayler**, who went out the way he would have wished (had he ever thought about it) and to Cecil Kimber who most certainly did not." Racing mechanic Tayler was Reg Jackson's *qv* chargehand, who died in Kaye Don's K3 Magnette (K.3021) during unofficial practice prior to the 1934 Mannin Beg race in the Isle of Man. It marked the end of Don's racing career, following a conviction for manslaughter.

It is well-known that MG's **telephone number** of Abingdon 251 provided the starting point for most MG chassis numbers, but the practice began prior to the move to the town. The first MG model to feature this prefix was the Morris Flatnose-based 14/40 Mark IV that entered production in November 1927 with the chassis number 2251, although the next car, the Mark I Six, began, appropriately, with 6251. But when the Mark II entered production at Abingdon in March 1930, the first chassis number was 0251, so beginning a trend that endured until 1953. Production TFs and Magnette ZAs used a 501 start, which was a Nuffield system that originated with Wolseley, and the phone number of its Birmingham factory was 1*501*! Having said that, not every MG built during the years 1930 to 1953 used the 0251 code, namely the J3, J4, K2, K3 and NE models.

In these circumstances it seems likely that Kimber requested the

251 number and MG was also allotted Abingdon 252 and 253. These remained until 1974, when Abingdon's manual exchange was one of the last in the country to switch to automatic use. MG's new number was Abingdon 25251.

The practice of beginning MG VIN (vehicle identifiable numbers) with 251 was revived by the Rover Group *qv* at the suggestion of the MG Car Club *qv* when the MG RV8 entered production in 1993, and the practice was perpetuated on the *MGF* of 1995.

Cecil Kimber's sudden departure from MG in November 1941 followed a meeting with **William Miles Webster Thomas** (1897-1980) in his capacity as vice-chairman and managing director of the Nuffield Organisation *qv*. Born in Ruabon, Wales, after education at Bromsgrove School, he served an engineering apprenticeship with Bellis and Morcom, of Birmingham.

Following war service in the Armoured Car Squadron in Africa and the Royal Flying Corps in Egypt and the Middle East, in 1918 he was awarded the Distinguished Flying Cross.

In 1919 Thomas joined the staff of *The Motor*. He had actually applied for a job on *Motor Cycling*, but only remained there for three hours before being switched to the motoring title as technical editor. There he was remembered for his astonishing memory and immense capacity for work.

Soon afterwards, in 1922, he took over as editor of *The Light Car and Cyclecar*, but by this time he had met William Morris *qv*, who was clearly impressed and, at the end of 1923, Thomas joined Morris Motors as a publicity adviser. In this capacity he became founding editor of *The Morris Owner qv* magazine and in 1927 was appointed a director and sales manager of Morris Motors *qv*.

Following Leonard Lord's arrival at Cowley, Thomas diplomatically moved to Morris Commercial, in Birmingham, as director and general manager, and transferred to Wolseley in that capacity in 1935. He became its managing director in 1937 following Oliver Boden's promotion to take up the vice-chairmanship of Morris Motors.

Thomas took over on his death in 1940 of what he renamed The Nuffield Organisation, and at the instigation of his chairman was responsible for sacking Kimber, obstensibly because he had seized the initiative and secured a contract for MG to build the centre-section of the Albemarle bomber. Sadly, the MG managing director suffered like so many of his talented colleagues who occasioned Lord Nuffield's displeasure. Just seven years later to the month, in November 1947, Thomas himself departed, and in 1949 became chairman of British Overseas Airways Corporation. One of the few Morris executives of international stature, in 1956 he became chairman of the British arm of Monsanto Chemicals before his retirement in 1963 and he became a life peer, Lord Thomas of Remenham, in 1971.

In 1964, after the death of Lord Nuffield, his autobiography *Out on a Wing* was published that revealed for the first time the circumstances of Kimber's dismissal. This was not Thomas' first book. Donning his journalist hat, in 1945 he had part-written the Organisation's wartime achievements in *Calling All Arms*, under the pseudonym of Ernest Fairfax.

To George Eyston *qv*, **John Thornley** was "a delightful and congenial human dynamo". When BMC competition manager designate Marcus Chambers *qv* visited Abingdon in 1954 he recalled in his book *Seven Year Twitch* that Thornley's enthusiasm was "infectious. When he first showed me round the works, we strode along with such speed that I almost had to trot to keep up with him, and I was treated to a running commentary as we went."

Appointed by Cecil Kimber in 1931, Thornley made the greatest contribution to the enduring success of MG, after its founder. Articulate, diplomatic and extrovert, with a lively sense of humour, he was perhaps the antithesis of the British Motor Corporation executive of his day.

Like Kimber, John William Yates Thornley's background did not suggest a career in the British motor industry. Born on June 11, 1909, the son of a master tailor, he was educated at Ardingly College and seemed destined for chartered accountancy; he was articled to the prestigious City of London firm of Peat, Marwick and Mitchell and, also like Kimber, studied the subject at evening classes, although he attended his at London University.

But it was his parent's divorce, when he was two, that indirectly changed the course of his life. Money, placed in a trust, which he received on his 21st birthday, was spent in June 1930 on a standard blue M-type Midget bought from Jarvis of Wimbledon *qv*, and this led to the MG Car Club *qv* and so to Abingdon. He joined in November 1931 as assistant service manager and became service manager, where his communication skills were used to maximum effect, in 1933. After war service in the Royal Army Ordnance Corps, where he rose to the rank of Lieutenant Colonel, he returned to Abingdon and in 1947 was appointed sales and service manager. The following year came promotion as assistant to general manager Jack Tatlow *qv*. In November 1952, 21 years to the day since he joined MG, Thornley was appointed general manager, a position he held until July 1969.

During this period MG embarked on an unparalleled era of growth, from 250 T-types being produced every week to some 1,300 MGs and

John Thornley, photographed soon after he became MG's general manager in 1953.

Austin-Healeys leaving the plant in the 1960s.

Abingdon's relative isolation from BMC's mainstream activities allowed Thornley to practice what he later described as design by *fait accompli*. "Why I didn't get kicked all the way from here to Birmingham and back I cannot imagine, but I did this constantly. There were major confrontations as a result, but no recriminations for having done so. By the time it was done they were always delighted."

Occasionally he did overstep the mark. One instance was the subversively conceived 1959 Le Mans car EX 186 *qv* that had to be hastily dispatched out of harm's way to America.

John Thornley was one of the very few managers to successfully establish a fruitful dialogue with BMC's dangerous and difficult chairman Leonard Lord *qv*, a relationship that thankfully ensured MG's, and Abingdon's, survival.

Sadly he succumbed to serious illness in September 1966, and although much later he was back in harness, he immediately found himself at odds with the new British Leyland administration of 1968, particularly Austin Morris' managing director, George Turnbull, former general manager of Standard Triumph. He retired, at the age of 60, in 1969.

Thornley's *Maintaining the Breed* appeared in 1950, and a second edition followed in 1956, which was reissued by Motor Racing

Publications in 1990, with 251 copies, MG's famous phone number, being donated to the MG Car Club *qv*. Thornley's home number was, incidentally, Abingdon 261!

John Thornley died on July 15, 1994, at the age of 85, but in 1970 he was inadvertently killed off by the now defunct *Veteran and Vintage* magazine that published his obituary in its February issue. In apologizing to him the following month, the magazine wished him "a long and happy retirement", which lasted for a further 24 years!

The Three Musketeers was the name given, in 1935, to a factory-supported team of NE Magnettes, finished like the initially privately run Cream Crackers *qv*, with cream bodywork with chocolate wings. Created for sporting trials, they carried the names of Athos (JB 4608), Porthos (JB 4750) and Aramis (JB 4606) inspired by the Dumas novels.

The later 1935 cars were the three P-types that had hitherto formed Eyston's Dancing Daughters *qv* team, although the 1936-season MGs were very different. These were based on the L-type Magna chassis, with the N-type 'six' bored out to 1,408cc; they were registered JB 6865, JB 6866 and JB 6867.

In 1937 the Musketeer's livery changed to red, and its TAs, registered ABL 961, ABL 963 and ABL 965, were fitted with supercharged engines. The 1938 team, BBL 82, BBL 83 and BBL 84, were Marshall-blown TAs.

The Musketeers survived into 1939, although the Cream Crackers were disbanded at the beginning of the year and, with hostilities looming, MG's official involvement in sporting trials came to an end.

The **Tickford** name has been associated with MG for over 50 years. Produced by Salmons and Sons, established by Joseph Salmons in 1820, the business remained in the control of this family until its sale in 1942. Established as a coachbuilding business in a cottage and outbuildings in the walled estate of Tickford Abbey, in Newport Pagnell, Buckinghamshire, its first car body dated from 1907. The Tickford name was first applied to Salmons coachwork in 1925 to describe the Tickford All Weather body that simplified the process of lowering the folding roof of a landaulette by mechanical rather than manual means.

Salmons made a speciality of drophead coupes, although these not always incorporating the Tickford system, and its first major contract with MG came in 1936 with the arrival of the Tickford body on the SA chassis. The option was later extended to the VA and WA. By far the most popular was the version used on the TA and a total of 260 were produced. A further 60 were allocated to the TB.

By the late 1930s Salmons was producing a sufficiently large number of bodies for Abingdon that it built them in what was called the 'MG Shop', identified by its octagonal badge mounted high on the wall. Chassis, complete with wings and a rudimentary body, were delivered from Abingdon, some 45 miles away, by road. When a representative of *The Sports Car* visited the works in 1938 he found that there were "some 500 people on the payroll, but the offices are in the house which was occupied by the great grandfather of some of the present heads of the concern". Because of the numbers involved the MG bodies were assembled in jigs.

Production ceased with the outbreak of war, and during hostilities, in 1942, Lucas and George Salmons sold the business to wealthy enthusiast Ian Boswell, who renamed the firm Tickford Ltd.

The VA and WA did not survive the war, and the drophead option was not perpetuated on the T Series cars. When Tickford did restart body production in 1947, it was on the Alvis TA chassis, and in 1952 it produced the body for the prototype Healey 100.

By this time Tickford was producing the Lagonda saloon body, but when Standard-Triumph secured the exclusive output of Mulliners of Birmingham, which built Aston Martin bodies, David Brown bought Tickford in 1955 to secure its continuity of supply.

Aston Martin production was transferred from Feltham to the former Tickford factory and later, in 1972, Brown sold out to Company Developments, but this went into receivership in 1974. Thankfully, the business survived and more changes of ownership culminated in Aston Martin being owned in 1981 by Pace Petroleum and CH Industrials.

The established Tickford name was reborn that year as Aston Martin Tickford, an independent automotive engineering business able to undertake work from companies other than Aston Martin.

In 1983 it moved to purpose-designed premises in nearby Milton Keynes, and the company's association with MG was then revived with the Tickford version of the MG Metro. At a cost of £7,505, that was £2,706 more than for the standard car; it was instantly identifiable by its front spoiler and redesigned bumpers. A leather dashboard featured inside and electric windows were standardized. Leather upholstery was available at extra cost. But this latter-day Tickford MG failed to catch on.

Later, in 1984, came a complete separation from Aston Martin, which was dropped from the company's name and ownership passed to the CH Industrials group. The link with MG was again revived when BL awarded it a contract for the completion of the low-production MG Maestro Turbo *qv* of 1985. This work was undertaken at Tickford's premises at Bedworth, near Coventry, the cars being transported there from Cowley. After completion, they were then returned for final inspection prior to dispatch.

However, in 1991, CH Industrials, which also owned Motor Panels, collapsed and Tickford was the subject of a management buy-out. It is currently engaged in the design and development of low-emissions, high specific output engines and the creation and production of prototype and niche market vehicles.

George Charles Tuck (born 1909), was MG's publicity manager from 1931 until 1939. He was born in Cambridge, but his family subsequently moved to Oxford and George was educated at Oxford High School. In the late 1920s he was working in the motor department of Gammage's, but he was a family friend of MG's sales and publicity manager Ted Colegrave. When the latter was looking for an assistant, he arranged with George to be interviewed by Cecil Kimber.

George Tuck

He was appointed publicity manager in 1931, whereupon Colegrave concentrated on sales, and Tuck went on to coin the company's famous Safety Fast *qv* slogan. He remained at MG until 1939, when he left to join the army, where he served as an ordnance officer.

In 1945 he became near-east representative of the Nuffield Organisation *qv* and sales manager of Nuffield Exports in 1949. Following the creation of BMC *qv*, George Tuck became deputy director of export sales and joined the South and Central Africa subsidiary board in 1957. He later made his home in South Africa, is a vice-president of the MG Car Club *qv* and still lives there at the time of writing (1998).

MG built no **twin overhead camshaft**-engined cars before the war and relied, instead, on a single-cam layout. But such a cylinder head was designed at Abingdon in 1932, although never made because of H N Charles' *qv* concerns about flame travel. However, Norton motorcycles did make a twin-cam head for MG from aluminium bronze, beautifully engineered with needle-bearing rockers. Paradoxically, the Coventry firm never achieved the power achieved by Abingdon without the expense of twin camshafts.

A conversion was effected on three R-type single-seaters *qv* in 1936 by Michael McEvoy, who had been involved with Laurence Pomeroy *qv* in the development of the Zoller-supercharged Q and R-type single-cam racers. He approached the Evans brothers, of the Bellevue Garage, on London's Wandsworth Common, with a view to designing the conversion as a challenge to the twin-cam single-seater Austins that were due to appear that year.

The resulting cast-iron heads, with valves at 90 degrees to one another, used the existing vertical drive with the cams driven by bevels and spur gears. The Zoller was mounted on the nearside of the engine instead of at the front, and the twin-cam 'four' gave 130bhp.

The cars, driven by Kenneth Evans, Ian Connell and Douglas Briault, first appeared at the British Empire Trophy race Donington in 1936 and were identifiable by their twin exhaust pipes on the offside, as opposed to the original nearside layout. But none finished and the heavier and more complicated 'double-wiper R-types', as they were known, thereafter enjoyed little success and soon withdrew from frontline competition.

This was despite the fact that one of these engines, running at 25lb boost, had been coaxed to develop 141bhp at 7,300rpm, although this was only marginally more than the factory achieved in 1936 on The Magic Midget's single-cam Q-type unit . . .

More successful was the unique McEvoy twin-cam head used on Reg Parnell's six-cylinder K3 (K.3009) that he raced in 1937, when it also acquired a single-seater body and ran in all the major meetings at Brooklands, Donington, Crystal Palace and the Isle of Man. Happily, the head has survived in the engine of the K1-based single-seater completed in 1984 for Roger Sweet.

U

University Motors, which was established in 1911, became MG's sole London distributors in 1930 and remained so until 1968. It is best remembered for selling its cars enhanced by an MG-prefixed registration number *qv*.

The company's chairman and managing director, Major George Bradstock (1888-1965), had a distinctly non-motoring background. Born in Cambridge, he was the younger son of the Rev J R Bradstock, educated at Blundells and Jesus College, Cambridge, and served in the Royal Artillery in the First World War, when he was awarded the DSO and Military Cross and bar.

He became managing director in 1921, and the company was clearly aiming at a student and post-graduate clientele like himself because its advertisements featured, without explanation, the coats of arms of Cambridge and Oxford colleges. Graduates of the former establishment, incidentally, often described Oxonians as having been educated in "the Latin quarter of Morris Cowley".

By the time it acquired the MG franchise, University had become very much of a family business. John Bradstock arrived in 1928 and became works director. Later, in 1949, George Bradstock's son Michael, educated at Eton College, joined the company and was soon appointed general manager.

Major George Bradstock

The association with MG had been well-timed and University Motors grew with the business. Earl Howe was a director, racing driver Hugh Hamilton a salesman and sales director Stanley Kemball chairman of the MG Car Club *qv*.

In the early 1930s its head office was at 7 Hertford Street, Mayfair, but its best-known London showrooms, both before and after the war, was at Stratton House, 80 Piccadilly. There was also a branch at

24 Bedford Place, Brighton.

The firm also offered 'University' bodies, initially the foursome drophead coupe on the M-type chassis that was also extended to the J2, Magna L, Magnette K2 and F-type. However, they were built by Carlton of Willesden, and Abbey at Acton. In 1946 the company bought Coachcraft, of Hanwell, which was renamed University Coachworks and subsequently moved to Egham to concentrate on car repairs. The Hanwell works was responsible for commercial vehicle bodywork and distribution.

By the time the firm marketed its MGC University Motors Special *qv* in 1969, Michael Bradstock was chairman of a business that had evolved into UMECO Holdings. By the 1970s its subsidiaries included University Electrics and University Marine.

The deep malaise affecting the British motor industry, coupled with the recession, took its toll on the business, which closed in 1986. However, University Motors was listed as a dormant company within UMECO of Hungerford, Berkshire, until the early 1990s.

V

The 1,548cc **VA** was the second of the pushrod-engined MGs, produced between May 1937 and September 1939, chassis numbers running from VA0251 to VA2657. A total of 2,407 were produced, mostly in saloon form, but there was an open four-seater tourer by Charlesworth and a Tickford drophead coupe.

The 2,991cc **V6 4V** engine, for V6 four valves per cylinder, used in the Metro 6R4 *qv* rally car, is unique in MG's history. This is because it is the only power unit to have been specifically designed for its purpose, as opposed to one that had been modified from an existing engine. It differed from many of its Group B rivals in not being turbocharged, this being offset by a larger cubic capacity. There was no suitable engine within BL at the time, the nearest being the Rover V8 *qv*, but it was too large, so it was cut down as a V6, with the resulting 2,495cc bringing it neatly within the 2.5-litre class limit. The original 1983 cars were so powered, but the definitive 6R4 did not appear until 1985, by which time it used this purpose-designed V6, the work of Austin Rover's David Wood. The 90-degree unit, with twin belt-driven overhead camshafts and four valves per cylinder, was built at BL's Radford works, originally a Daimler shadow factory that Standard-Triumph had taken over in 1957. Sadly, the engine, which developed 380 or 410bhp, depending on its tune, never achieved its expectations, and the 6R4 failed to win a major event.

Group B, for which it had been designed, was axed in 1987 and the engine was bought by Tom Walkinshaw in 1988. He heavily reworked the V6 and introduced a desirable element of reliability for it to power his XJR-11 sports-racer of 1990 and, ultimately, the Jaguar XJ220 supercar of 1992.

The **V8 engine** that powered the MGB GT V8 of 1973-76 and the 1992-95 MG RV8 originally appeared in the 1961 Buick Special compact range. Announced as the 'Aluminium Fireball V8', it was a small unit by transatlantic standards, with a capacity of 215.5cid

(3,531cc). But the really significant aspect of the design was that it was the first American V8 to be mass-produced in aluminium alloy.

For Buick's General Motors parent the incentive was to reduce overheads by manufacturing an engine in which the more costly alloy block was of sufficient hardness and wear resistance for the pistons to run directly in it. This would obviate the need to fit cast-iron cylinder liners, so cheapening and simplifying the production process.

GM had road-tested experimental aluminium V8s as early as 1952 and, after six years, the project was handed over to Buick. It strove to eliminate the use of liners by increasing the silicon content of the aluminium alloy. But despite promising results it was unable to prevent the piston rings scuffing the bores from a cold start, even though aluminium warms up quicker than its cast-iron equivalent. This did not distract from the fact that aluminium-to-aluminium contact was perfectly satisfactory at continuous high speeds.

As a result the Buick engineering team, headed by Joe Turlay, decided on a compromise by reducing the silicon content of the aluminium and introducing cast-iron cylinder liners to the engine. This pushrod over-square unit with 3.5in (88.9mm) bore and 2.8in (71mm) stroke, entered production in this form. It developed 155bhp at 4,600rpm. The block was also used by Buick's Oldsmobile stablemate, but with orthodox wedge-type combustion chambers.

However, the engine was destined for a relatively short manufacturing life and was discontinued at the end of the 1963 model year. One of the factors in its demise was that it took no less than four gallons of coolant and required a special antifreeze. Using the incorrect type would release aluminium-silicone oxide, so clogging the radiator and prompting cylinder head warping.

The V8's relatively low cubic capacity was not a factor in its demise because the design and tooling was then applied to a 198cid (3,244cc) cast-iron V6 that Buick used from the 1962 season until 1967.

So after the production of some 750,000 units, the V8 was discontinued. But the alloy unit was to prove its worth in Australia when, reduced to 3 litres capacity and converted to a single overhead camshaft per bank, it powered the Formula One Repco-Brabham that took Jack Brabham to the drivers' World Championship in 1966.

In the meantime, early in 1964 William Martin-Hurst, managing director of the still independent Rover company, had been in America and discussed with J Bruce McWilliams, head of the company's North American operations, the latter's revolutionary idea of using a small American V8 in Rover's car range.

Later that year, on a visit to Karl Keikhaefer, president of GM's Mercury Marine, to whom Rover was considering selling engines, Martin-Hurst spied an obsolete Buick unit . . . A deal was effected in

January 1965 and, despite considerable resistance to the idea at Solihull, the V8 duly first appeared under the bonnet of the ageing 3-litre-based 3.5 saloon of 1967.

In taking over the manufacturing rights, Rover was able to make some modest changes to the block, the original gravity casting process being replaced by a sand casting. Only one American component remained in the engine, the Diesel Equipment Company of Grand Rapids' hydraulic tappets. The original four-jet single-barrel carburettor was replaced by twin SU HS6 carburettors. The resulting engine developed 184bhp at 5,200rpm.

The V8 subsequently found its way under the bonnet of the 2000-based 3500 of 1968-1976, powered the Range Rover (1970), the Rover SD1 hatchback (1976-86) and the Land Rover (1979).

Its sporting potential had immediately been identified by Morgan, and its Plus 8 (1968) was so powered, and in 1969 by Ken Costello *qv*, which led to the creation of the MGB GT V8 of 1973. However, the Abingdon-built version used the lower-compression Range Rover engine, and this produced problems for MG because it continually suffered from a shortage of engines.

This was a closed car, but it did not prevent enthusiasts from introducing the unit in their own roadsters, those by Bertie Samuelson and Roger Parker being cases in point. The latter car was of significance because it showed the viability of the concept that flowered as the MG RV8 in 1992.

By this time the long-running 3,528cc unit had briefly appeared in the TR7 in rally and TR8 forms in 1980, and been enlarged in 1989 to 3,946cc by increasing the bore size to 94mm. Fuel injection had been available since 1982, although the Lucas L electronically controlled system had been applied to V8 Rovers in the Australian market since 1977.

An increase to a 93.5mm-bore 3,905cc had already been undertaken in 1984 by TVR, the engine having been supplied by Rover for use in its Tasmin model since the previous year. But it was Rover's 3.9-litre engine that was used in the TVR Griffith of 1991 and in the 1993 Chimaera. It was essentially unchanged for the MG RV8 of 1992-95, which developed 187bhp at 4,760rpm.

The V8 is still used in 3.9-litre form in the Morgan Plus 8 and in heavily revised 4.0 and 4.6-litre guises in the current Range Rover and Land Rover Discovery models.

The most luxurious MG of its day, the 2,561cc six-cylinder **WA**, successor to the SA, was mostly produced in saloon form. A total of 369 were built between October 1938 and September 1939. Chassis numbers ran from WA0251 to WA0619.

When Fred Mouldy became one of MG's few Abingdon-based apprentices in 1930, his **wages** were 11s (55p) a week and he had to provide his own tools. Jimmy Simpson, who entered the company's sales office at about the same time, received 12s 6d (62p), whilst tea boy Dutchy Holland earned 11s 3d (56p). Working hours were 8am to 6pm on weekdays and to 12 noon on Saturdays, and in these prewar days there were layoffs, sometimes for months at a time.

One of Britain's oldest car companies, **Wolseley Motors** was responsible for the initial design of the four and six-cylinder overhead-camshaft engines used in the Midget, Magna and Magnette models produced between 1928 and 1936. Building its first car in 1896, Wolseley was bought in 1901 by the Vickers armaments group, but by 1914 it was the British motor industry's most productive but inefficient car company.

During the First World War, its factory at Adderley Park, Birmingham, produced the single-overhead-camshaft V8 Type A Hispano-Suiza aero engine from which evolved the Wolseley W4 Viper. The power units of its postwar car range perpetuated this mechanically efficient but costly feature. These were produced at a new factory built at Drews Lane, Ward End, Birmingham, for the manufacture of its Stellite car. But Wolseley continued to be cash hungry, and in 1926 Vickers put the business up for sale and William Morris *qv* secured it in the face of strong opposition from his great rival Sir Herbert Austin, who had at one time been Wolseley's general manager. Much of the £730,000 that Morris paid came from the successful public flotation of Morris Motors *qv* in 1926. The Adderley Park works became, in 1933, the home of Morris Commercial Cars.

Wolseley's Ward End works was responsible for the design of the 53 x 83mm 847cc four-cylinder engine, with its distinctive up-ended dynamo doing double duty as the drive to the single overhead camshaft. The layout was patented by Wolseley and used in the Morris Minor of 1928. Just as he had with Morris' Engines Branch *qv* JA unit, Cecil Kimber immediately recognized its performance potential and it was extended to the MG M-type Midget. A 1,271cc six-cylinder version, sharing the same bore and stroke, first appeared in Wolseley's undistinguished Hornet model of 1930, and it was extended to MG's F-type Magna of 1931. But from thereon Wolseley and MG engine design diverged as Ward End dispensed with the vertical dynamo and replaced it with a cheaper, less positive chain. This permitted the firm to move the engine further forward in the chassis of the Hornet Special.

Hubert Charles' small team at Abingdon continued to modify the power units that became more MG than Wolseley. He liaised closely with A V Oak, who had joined the firm in 1907 and with whom he later worked when Oak moved to Cowley in 1936.

Abingdon introduced the cross AA head in 1932 and the three-bearing crankshaft engine used in the P-type Midget and Q and R-type racing cars were peculiar to MG. But when Morris Motors took over at Abingdon in 1935, Ward End's design and manufacture ceased, and thereafter MG's, and indeed Wolseley's power units were produced at the Coventry premise of Engines Branch.

The Nuffield Organisation's *qv* manufacturing operations were restructured in 1949, Wolseley car production was moved from Birmingham to Cowley, and the make endured until 1975. This meant that the former car factory at Ward End could become a machining centre for the entire business. The Nuffield Universal Tractor also entered production there in 1948, and the facility was renamed the Tractors and Transmissions branch. As such it was responsible for producing the hub and brake assemblies for MG sports cars as well as the rear axle for the MGA and MGB. The works survives today as the home of LDV commercial vehicles.

Frank George Woollard (1883-1957) was responsible with George Arthur Pendrell *qv* for the JA-type 2.5-litre six-cylinder single-overhead-camshaft engine used in the MG 18/80 *qv*. A production engineer of international standing, bearded and bespectacled, London-born Woollard was educated at the City of London School and Birkbeck College, and served an engineering apprenticeship with the London and South Western Railway.

Woollard's first connection with the motor industry came in 1907 when he joined the staff of Weigel Motors, in London, but he moved

Morris Motors' talented production engineer Frank Woollard.

in 1910 to the Birmingham axle manufacturer E G Wrigley, where he was chief draughtsman. The company supplied the front and rear axles and the steering box for the Morris Oxford of 1913.

Woollard later became Wrigley's chief designer and ultimately assistant managing director. It was there that he met Cecil Kimber *qv*, who worked for the firm from 1918 until 1921. When Wrigley got into financial difficulties, Woollard left in 1923, but Morris sought him out and it was on his advice that the latter bought Hotchkiss' Coventry factory, which was renamed Morris Engines. He asked Woollard to manage it and he was responsible for laying down an innovative system of mass-producing engines by automated transfer machines, which was regarded as the most advanced process in the world, inclusive of America.

Frank Woollard remained with Engines Branch until 1932, when he left following a disagreement with Morris, one of a number of the firm's talented engineers to do so. When he departed he joined Rudge Whitworth and became its manufacturing director. In 1936 he resigned to take up a position with Birmingham Aluminium Casting and its associated company Midland Motor Cylinder. He left in 1949 to become a consultant and was president of the Institution of Automobile Engineers in 1945-46 and 1946-47 and spent his spare time reading, writing and "studying human relations in industry". His book, *The Principles of Mass and Flow Production*, was published in 1954. Woollard died in December 1957 at the age of 78.

X

Morris' Engines Branch designed the **XPAG** power unit that first appeared in the 1939 TB sports car and went on to power its TC successor, the Y-type saloon, the TD and the TF, then ceased production in 1956. A successor to the MPJG *qv* engine used in the TA, it first appeared in XPJM form in the Morris Ten, announced in August 1938. Unlike the earlier Ten, it was an overhead-valve unit and, instead of the enduring 102mm stroke, it had a shorter 90mm throw whilst retaining the 63.5mm bore, which gave 1,140cc.

For the TB that appeared in April 1939, the bore size was increased to 66.5mm, giving a capacity of 1,250cc, and it developed 54bhp at 5,200rpm. It was instantly identifiable by its air filter, mounted at right-angles across the rocker box.

MG's robust Morris-based XPAG engine seen here under the bonnet of a Y-type saloon.

The XPAG engine had a strong, robust block and, unlike its MPJG predecessor, this met the sump at the centreline of the crankshaft. The opportunity was also taken to dispense with the cork-faced clutch and replace it with a conventional proprietary Borg and Beck single-plate unit.

The pushrod 'four' was carried over in virtually unchanged state for the TC and TD, although for the TD II that became available to special order between 1951 and 1953, the compression ratio was raised from 7.25 to 8:1, thus increasing output to 60bhp at 5,500rpm.

The XPAG engine in the TF was modestly tuned to give 57bhp, but what was really needed was an increased capacity to 1,500cc. Amateur racers had succeeded in enlarging the engine's capacity, but it was a tricky business with the attendant risk of breaking through into the water jacket and new cylinders liners were often needed.

Morris' Engines Branch *qv* recored the block, which permitted the bore size to be increased from 66.5 to 72mm, giving a capacity of 1,466cc. This first appeared in unsupercharged form in the EX 179 *qv* record-breaker of 1954, and at the end of the year it was extended to the retitled TF 1500, which had its top speed increased from 80 to 85mph and some 2.5 seconds shaved off the 0-60mph acceleration time.

But this redesignated XPEG unit was only destined for a brief life until the TF ceased production in April 1955, although it continued in its original single-carburettor XPJW 1,250cc form in the Wolseley 4/44 saloon until mid-1956.

The **XPAG** engine in its formative 1,250cc and later 1,466cc states powered a host of MG Specials of the late 1940s and early 1950s, most notably by Cooper, Kieft, Lester, Leonard and Lotus.

Y

The 1,250cc **Y-type** saloon was conceived at Cowley before the war, Morris Eight-based, but Abingdon-built. Designated EX 166, it was produced between April 1947 and August 1951. By this time 6,158 examples had been made. Chassis numbers ran between Y0251 and 7020. The YT touring version, available between October 1948 and August 1951, accounted for 877 cars and ran between YT1885 and 7020. The Y was succeeded by the YB, which was available between January 1952 and September 1953, this accounting for a total of 1,302, with numbers running from YB0427.

Only a few **Y-types** were fitted with special bodies. Just 11 chassis were supplied, nine in 1948, and one each in 1952 and 1953. Swiss coachbuilders Reinbolt and Christe, of Basle, which had rebodied some SA saloons, ordered three Y-type chassis in 1948 and fitted touring bodies ahead of the factory's YT. Beutler, also Swiss-based, bodied at least one drophead coupe. Across the Alps in Milan, Zagato produced a one-off coupe in its Panoramica series introduced in 1947. A feature was curved window plexiglass, and a so enhanced Y-type was displayed by Zagato at the 1948 Italian motor show at Turin. This was the outcome of a commission from Roger Barlow, of International Motors of Los Angeles, who visited Abingdon, and there was talk of some 70 being produced. Farina and Castagna were also mentioned and roadsters and two and four-seater drophead coupes were planned. A Shorrock supercharger was to be fitted. But in the event only one further car was built, a Castagna coupe.

The ***Ye Olde Bell*** hotel at Hurley, Berkshire, was where the fate of MG's Abingdon factory was sealed. It was there that Michael Edwardes and his management team met in the summer of 1979 to decide the strategy of their Recovery Plan, announced on Black Monday *qv*, September 10.

Z

The 1,489cc **Magnette Z Series** saloon was the first MG, following the creation of BMC in 1952, to be powered by an Austin engine. It revived the Magnette name although, perversely, it was not a 'six'. Available between late 1953 and September 1956, 12,754 were built and chassis numbers ran from KA501 to 18576. Those of its ZB derivative, available until late 1958, ran from 18577 to 37100 and 23,846 were produced.

One Z Series Magnette driver takes over the sash in a six-hour relay race at Silverstone in the 1950s.

Bibliography

Books:

The Art of Abingdon, by John McLellan (Motor Racing Publications, 1982)
Aspects of Abingdon, by Marcham Rhoade (MG World, 1996)
The Story Behind the Octagon, by Brian J Moylan (Privately published, 1995)
Early MG, by P L Jennings (Privately published, 1989)
K3 Dossier, by M F Hawke (Magna Press, 1992)
The Kimber Centenary Book (The New England MG T Register, 1988)
The Magic of MG, by Mike Allison (Dalton Watson, 1972)
*Maintaining the Breed**, by John Thornley (Motor Racing Publications, 1956 and 1991)
Making MGs, by John Price Williams (Veloce Publishing, 1995)
*MG by McComb**, Wilson McComb (Osprey, 1984)
MG Collectibles, by Michael Ellman-Brown (Bay View Books, 1997)
An MG Experience, by Dick Jacobs (Transport Bookman, 1976)
MG Mania: 'The Insomnia Crew', by Henry W Stone as told to Dick Knudson (New England MG T Register, 1983)
MG Sports Cars, by Malcolm Green (CLB International, 1997)
MG: The Sports Car America Loved First, by Richard L Knudson (Motorcars Unlimited, 1975)
MG T Series: A Collector's Guide, by Graham Robson (Motor Racing Publications, 1980)
Original MG T Series, by Anders Ditlev Clausager (Bay View Books, 1991)
*MG The Untold Story**, by David Knowles (Windrow and Greene, 1997)
MG V8, by David Knowles (Windrow and Greene, 1994)
The Mighty MGs, by Graham Robson (David and Charles, 1982)
Original MGA, by Anders Ditlev Clausager (Bay View Books, 1993)
MGA, MGB and MGC: A Collector's Guide, by Graham Robson (Motor Racing Publications, 1977)

MGB The Illustrated History, by Jonathan Wood and Lionel Burrell (Haynes, 1988 and 1993)
Original MGB, by Anders Ditlev Clausager (Bay View Books, 1994)
*Oxford to Abingdon**, by R I Barraclough and P L Jennings (Myrtle Publishing, 1998)
Triple M Yearbooks (The MG Car Club T Register)
* Strongly recommended

Related Books:

A-Z of British Coachbuilders, by Nick Walker (Bay View Books, 1997)
The BMC/BL Competitions Department, by Bill Price (Haynes, 1989)
Bouverie Street to Bowling Green Lane, by A C Armstrong (Hodder and Stoughton, 1946)
The Bullnose Morris, by Lytton P Jarman and Robin Barraclough (Macdonald, 1965)
Calling All Arms, by Ernest Fairfax (Hutchinson, 1945)
The Guv'nor, by John Howlett (Privately published, 1973)
The Life of Lord Nuffield, by P W S Andrews and Elizabeth Brunner (Blackwell, 1955)
Magic MPH, by Lieut Col Goldie Gardner (Motor Racing Publications, 1951)
The Morris Motor Car 1913-1983, by Harry Edwards (Moorland Publishing, 1983)
Out on a Wing, by Miles Thomas (Michael Joseph, 1964)
Wheels of Misfortune: The Rise and Fall of the British Motor Industry, by Jonathan Wood (Sidgwick and Jackson, 1988)
Who's Who in the Motor Trade (Motor Commerce, 1934)
Who's Who in the Motor Industry (1952/1972)

MG Magazines:

Enjoying MG
MG enthusiast Magazine
The MG MaGazine
MG Magazine
MG World
Safety Fast
The Sports Car

Related Magazines, Newspapers and Periodicals:

Abingdon Herald
The Autocar (now Autocar)

Classic Cars
Classic and Sportscar
Proceedings of the Institution of Automobile Engineers
The Morris Owner
The Motor (later Motor)
Morris Register Magazine
Old Motor
Oxford Mail
Oxford Times
Veteran and Vintage Magazine